AF564822

2nd Revised and Enlarged Edition

Canopy Management of Fruits

NIPA® GENX ELECTRONIC RESOURCES & SOLUTIONS P. LTD.
New Delhi-110 034

About the Author

Dr. Kanchan Kumar Srivastava born in Sultanpur, U.P, obtained his B.Sc. (Ag) and M.Sc. Ag. (Horticulture Fruit Science) degree from Acharya Narendra Dev University of Agriculture and Technology, Faizabad, U.P. and Ph.D. (Hort- Fruit Science) from Institute of Agricultural Sciences, Banaras Hindu University, Varanasi, U.P. Dr. Srivastava is a recipient of President's Rajbhasba Gaurav Award (2015) for his original Scientific Book (Sheetoshn Phalon Ki Vaigynik Kheti, by DKMA-DIPA, ICAR) and awarded with UGC fellowship for Ph.D after attaining first rank in Research Entrance Test (RET) of BHU in 1998 and Junior Research. Fellowship by Indian Council of Agricultural Research, for M.Sc., in 1995, Dr. Srivastva awarded with fellow for the year 2017 by Indian Academy of Horticultural Sciences (IAHS) (Formerly known as Horticultural Society of India, New Delhi) and Fellow of 2017 by Horticultural Research and Development (SHRD), Ghaziabad, Uttar Pradesh, Young Scientist Associate Award (2012, SIEMAT, Bioved) and best DUS test centre award 2010-11.

Dr. Srivastava started his career as Assistant Professor at SKUAST-K, Srinagar, J&K in 2001 and joined as Senior Scientist in Fruit Science at ICAR-Central Institute of Temperate Horticulture, Srinagar in 2008 and joined as, Principal Scientist in 2014. Dr Srivastava's field of specialization is tree architecture management, high density planting, and Fruit Crop Improvement. He has been Principal Investigators of 30 external and internal funded projects. Dr Srivastava developed 04 apple hybrid (Ammol, Ammrit, Priam and Pride) Dr. Srivastava has released 04 sweet cherry, 05 walnuts, 01 apricot varieties and also collected good numbers of *Malus*, *Pyrus*, *Prunus* and *Ambri* apple genotypes, added in National Active Germplasm Sites Repository of ICAR-CITH Srinagar, J&K.

Dr Srivastava developed the efficient tree architecture techniques i.e. Espalier for Ultra High Density Planting System for apple and rejuvenation techniques for old and senile almond orchards. developed the DUS descriptors for testing of pome fruits (apple and pear) and stone fruits (sweet cherry and apricot). He has acted as Course Director for organizing 04 National Training Programmes for Scientists and Extension Scientists. During his carrier he has published more than 95 research papers in National and International Scientific journals, 03 Scientific books, 01 Monograph on Apple varieties of India and 20 extension bulletins in English, Hindi and Urdu.

2nd Revised and Enlarged Edition

Canopy Management of Fruits

Kanchan Kumar Srivastava
Ph.D. (Horticulture, Fruit Science), FIAHS, FSHRD
Principal Scientist
ICAR-Central Institute for Subtropical Horticulture
Lucknow-226 101, Uttar Pradesh

NIPA® GENX ELECTRONIC RESOURCES & SOLUTIONS P. LTD.
New Delhi-110 034

NIPA® GENX ELECTRONIC RESOURCES & SOLUTIONS P. LTD.

101,103, Vikas Surya Plaza, CU Block
L.S.C. Market, Pitam Pura, New Delhi-110 034
Ph : +91 11 27341616, 27341717, 27341718
E-mail: newindiapublishingagency@gmail.com
www: www.nipabooks.com

For customer assistance, please contact
Phone: + 91-11-27 34 17 17
Fax: + 91-11-27 34 16 16
E-Mail: feedbacks@nipabooks.com

ISBN: 978-93-90591-05-3

Composed and Designed by NIPA®.

Dedicated to
My beloved Mother and Papaji

Late Smt. Prema Devi Ji

(18 November 1944 to 7th April 2021)

and

Sri Dinkar Prakash

(1945 to 2nd May, 2004)

Preface

India ranks second in fruits with an annual production of 97.38 million metric tons, accounting for about 10.3% of the total world fruit production. However, the productivity of major fruits per hectare (14.97 tones/ha) is far below compared to world average (25 tones/ha). Among the various biotic and abiotic factors responsible for low yield and quality is a lack of knowledge about canopy management. Time bound removal of quantities curbs on import and other barriers to access to domestic market under World Trade Organization of which India is signatory, will require Indian horticultural produce to be competitive both in the domestic and export market. This would call for use of modern production technology i.e. high density orcharding, the mechanization of orcharding system and canopy manipulation and development. The light penetration and interception in canopy area can be improved by manipulating the canopy shape and size. Most of the old orchards are dense, over- crowded and their bearing is confined to upper periphery (tip bearing) of canopy. Further it harbors various pest and disease-causing agent due to change in micro-climate inside the canopy. The perfect canopy shape depends on the type of fruit plant and species. Keeping this in mind the present book has been prepared on various aspects of canopy management of fruit crops. This book covers all important fruits of temperate, tropical and sub-tropical and has been prepared keeping in mind the recent curriculum of Indian Council of Agricultural Research, New Delhi.

Though the utmost care has been taken to satisfy our esteemed readers, there may be some lapses which we hope will notice and aid us greatly benefit in improving chapters in the subsequent edition.

I take this opportunity to thanks Ms Shruti Srivastava for sketching diagram and also all my nears and dears for their constant encouragement for preparing the manuscript

Last but not least from the core of the heart I thank my beloved mother Late Smt Prema Devi and my Father Late Sri Dinkar Prakash for showering blessings, without their blessing it was not possible to make this manuscript. Thanks are also acknowledged to my better half Mrs. Shalu Srivastava, Aarnavi (daughter) and Lakshya (son) for whole hearted support.

Place: Lucknow
Date: July, 2022

Author

Contents

1

Introduction, Importance and Scope of Canopy Management

Over the last decade, area under horticultural crops grew by 2.60% per annum and production 4.8% per annum. During 2017-18 total 311.71 million tonnes horticultural crops were harvested from 25.43 million hectare area. India is 2nd largest producer of fruits in the world after China, with average productivity at 14.97 t/ha. Fruits alone were cultivated almost in 6.5 million hectare area and production of 97.38 million metric tons, (Anonymous, 2018). If compare with horticultural advance countries our productivity is quite low. The foremost reason of low productivity is poor orchard floor management, canopy management, and low density plantations. In mango most of the old plantations are not being provided proper water and nutrition with no or very little canopy management and practices, most of the old plantations have become overcrowded with least light interception, no or poor insect, pest and disease management. Therefore, average productivity of most of the fruits are quite low. Canopy management determind plant population per unit area which affect productivity. This chapter will deal with the importance and scope of canopy management practices.

Canopy management of fruit crops plays crucial role in orchard planning, fruit quality, productivity, tree density per unit area; all aforementioned factors affect total fruit production and productivity. Canopy management is one of the important operations in perennial fruit crops to harness the solar energy, utilizing space and soil nutrient and moisture.

There is currently a worldwide trend in fruit producing countries to accommodate more number of plants per unit area by manipulating tree spacing and growth using canopy management practices. To control the tree growth and shape, ultimately limiting tree size while still maintaining high fruit production of the desired quality.

Canopy in a fruit tree refers to its physical composition, comprising the stem, branches, shoots and leaves. The canopy is determined by the number and size of the leaves and its architecture is determined by the number, length and orientation of braches and shoots. Canopy management of the fruit trees deals with the development and maintenance of their structure in relation to the size and shape for maximum quality yield. Understanding the bearing behaviors of fruit tree species and influence of the environmental factors are required for canopy management. The basic principle of the canopy management of fruit crops is to make best use of the land and the natural resources for increased productivity.

Canopy physically supports the fruit and markedly influence the yield and quality. The canopy configuration largely depend on the orchard production system which is a combination of four variables i.e. (i) the variety (ii) the rootstock (iii) tree spacing and (iv) training system. Innovations in the production systems have witnessed significant variations in the tree canopy form in the last few years. The introduction of new production systems like bed system, Lincoln and Tatura trellis forms of canopy has brought significant transformation in the production system. Canopy management based on the choice of tree size, number of tree per hectare and pruning techniques, so as to optimize the production of assimilation and its conversion into economic yield.

Light is the most important component of fruit production, it has a major role in flower induction as well as in fruit development through carbohydrate synthesis. However, light interception and distribution within the tree canopy of tropical and subtropical fruit crops also influenced the process. The utmost objective is to find ways of maximizing light interception in the tree canopy through horticultural techniques, because light energy falling on the grasses in the alley-ways obviously is not producing fruits. The second aim is to optimize the light distribution within the canopy. In order to maximize the light use efficiency is maximized for the increased photosynthesis, fruit bud production, fruit growth and fruit coloring.

Importance of canopy management

Thorough knowledge on understanding of the canopy development and management is of paramount importance to achieve optimum efficiency of the orchard. The canopy intercepts light and manufacture carbohydrates used in growth and metabolic activity. Light distribution within the canopy affect fruit quality and color development. Canopy development influenced pest infestation and their control. Pesticides are expected to continue to play a major role for the foreseeable future in protection of fruit crops from insect and disease damage. Chemical approaches (Use of pesticides is important for pest and disease management), however, safeguarding the environment, is necessary in view of increasing chemicals cost etc. Orchard design make greater use of small trees in high density planting; change in tree forms as a result of new genetic material (e.g. dwarf and spur type) or cultural practices (pruning, training and canopy modifications by growth regulators) increased requirement for more precise delivery of predicted dose of the pesticide over the larger area. The change in tree size, shape and canopy geometry ease the problem of target oriented spray of pesticides and growth regulating chemicals. Use of spur bearing cultivars, central leader training system and pruning practices have resulted **better exposure of the** foliage and branch configurations within the tree, making spray penetration easier and deeper. In orchard system the target is to control the insect-pest and disease is complex phenomenon. The pest location within the canopy may be quite variable i.e. outside, inside and top of the canopy. Target-oriented protection of disease and pest measures have become important for precise spraying of pesticide for high quality fruit production.

Application of pesticides on targeted points and making adjustments in spray delivery protocols is a very important issue for the perennial nature fruit crop due to increasing spray chemical costs and government regulations for pesticide use. Chalking out of canopy management strategies that address these issues will also clearly pay dividends for the growers willing to invest in the management expertise to solve these problems.

Pruning, for the training and application of growth regulators to control vegetative growth are designed to improve the light penetration; thereby increasing the overall bearing potential and improving yield and fruit quality attrbutes. Hence, light interception becomes the main factor influencing tree productivity. Accordingly, model of the tree efficiency may best be expressed on the basis of yield per unit canopy light interception. Development of efficient canopy and computer aided models may help the horticulturist to simulate canopy design. Development of three-dimensional computer-aided design programme may be an approach whereby light distribution patterns can be evaluated efficiently for different canopy designs.

High yields of quality fruits are results of the combination of good light distribution in the canopy and high light interception. Further high tree density combined with thin canopy depth resulted high yields as well as high input use efficiency due to maximum light distribution in the canopy.

In USA, fruits harvesting cost accounts 17-30 per cent of the total production cost and for mechanical harvesting equipment developed in the early 1960's was largely the shake and catch methods, designed to remove fruits form low density planting in free standing orcharding systems. However, the mechanical harvesting results in fruits damage. Hence, limiting tree height to 2-3 m was suggested to facilitate harvesting and reduce fruit damage. Development of dwarf canopy model may helps to reduce the harvesting cost and time.

Scope of canopy management

Canopy architecture development is like laying of foundation for a huge building construction. Perennial nature fruit crops tends to attain maximum leaf area by mid-summer but usually intercept not more than 65 to 75% of available light at full canopy structure over the years. The together with the relatively low light interception at full canopy, this offers real scope for the Crop Physiologist, Pomologist and Plant Breeder to improve life-time orchard light interception.

Plant breeder, Pomologist, Growth Regulator Physiologist and Engineers must work together towards early achievement and easy maintenance of canopies designed to optimize light interception and distribution for advancing towards Golden Revolution. With the advent of WTO regime, currently canopy management has become one of the most exciting and prime area in fruit research.

In order to optimize the utilization of light for increased yield of quality fruits, canopy management deserves greater attention by exploiting the various horticultural techniques i.e. through the selection of proper rootstock and scion combinations, training and pruning systems, cultural practices, plant growth regulators, nutrient and water management practices and tree and branch orientation etc which will helps in maintaining the ideal canopies of the trees.

Tree with vigar and dense canopies used to be low productive, poor fruit color development and suffer various disease and pests. In case canopy height increased there is scope to increase the productivity per unit area. Further, the bearing behaviors of tree, convenience in cultural operations restricted the height of the tree in limit, forcing to develop dwarf and spreading canopies.

With increased in demand for quality fruits and to fulfill the nutritional demand of burgeoning population, it is the immediate need to accommodate almost 10 times more plant per unit area through high density planting system. This is possible only with the management of canopy to intercept maximum light in canopy most part of the tree. The canopy management has direct correlation with dry matter production, flower bud initiation and fruit quality. From the research it has been ascertained that the canopy management through advance training systems in fruit plants has been found more effective. The canopy model also influenced the water use efficiency, though it is not of significant amount. Under varied canopy architecture systems, the difference in canopy water use were not large, but yield differences were considerable (34 t ha^{-1} for the Vase Canopy and for Tatura Trellis Canopy 2 ton/ha). The water use efficiency was highest on those canopies giving the high yield. The Tatura Trellis used more water than the other system (Chootummatat *et al*., 1990).

Canopy management has significant role in controlling the tree growth, well developed canopy accomodate more tree per unit area. Canopy management practices has also been following in vegetables, flowers, spices, medicinal and aromatic crops and also in agronomic crops apart from fruit crops.

Tree architecture development and management have resulted accommodation of more tree per unit area under different fruit crops i.e. apple, pear, peach, plum, almond, mango, guava, pomegranate, pineapple etc. If the tree canopy management techniques are adopted, productivity under fruit crops can be increased many fold.

2

Principles of Canopy Management

Tree canopy management affects fruit yield and quality by affecting light interception. The canopy geometry should be managed in such a way that it intercepts maximum light. For high density planting system a shallow (thin) canopy 1.5-2.0 m in depth is needed to achieve the maximum efficiency for trapping sun energy through foliage and channelizing metabolites for quality fruit production with good return. The objective of the canopy management lies in the fact, as to how best we manipulate the tree vigor and use of the maximum available light and temperature to increase productivity and quality and minimize the adverse effect of the weather.

Light is an important aspect of canopy studies, because of it's role in photosynthesis, and its need in development of morphology of leaves and shoots, role in flower initiation, fruit-set, yield and quality. Light is the fundamental aspect of the fruit production. The objective of canopy management is to optimize the plant model in such form so that it may intercept maximum light by tree training, pruning, branch and tree orientation, use of growth regulating chemicals, planting space and design. Two aspects i.e. light and tree canopy development must be considered. First, interception of light by total orchard canopy is correlated to the total biomass production of the crop, Monteith (1977), reported that light penetration and distribution within a single tree canopy fruit must be optimized for harnessing the tree genetic potential with regard to yield and quality attributes.

The basic principles of canopy management as described by Shikhamany (2001) are (i) Maximum utilization of light (ii) Avoidance of the build-up of micro- climate congenial for growth of disease and pest (iii) convenience in carrying out the cultural operations (iv)Maximizing the productivity and quality (v) Economy in obtaining the required canopy architecture. As per report in many fruit crop, 15-20% increased in production of quality fruits have been noted by managing the appropriate and efficient canopy architecture in short statured tree (Nath and Pongener, 2017).

Significance of canopy size

Tree canopy size plays important role in orchard management and production of quality yield. Dwarf tree stature and short canopy have many added advantage. In dwarf canopy, solar radiation has good penetration which favours photosynthesis activity, thus tree produce high and better quality fruits. The entire tree is close to main trunk resulting very effective spray penetration which reduced in use of chemicals. Harvesting of the fruits from well managed canopy and pre harvest fruit bagging is easy, which gives a greater management efficiency to the orchardist.

Four components determine the tree dwarfness i.e. inter-node length, branching angle, branching location (basitonic, mesotonic and acrotonic), rate of tree growth. Besides the genetic constituents many horticultural techniques/ manipulations i.e. grafting, use of dwarf scion and rootstock cultivers, withholding of water (creating stress), close planting system, training the branches in horizontal position etc. determine tree dwarfness.

Light interception in orchard

Orchard is a discontinuous canopies, usually more light transmitted to the ground if tree height is low and the trees are distantly planted. Hence, light interception of the orchard can be enhanced by close planting the tree and increasing the tree height. Light transmittance through the canopies depends on the leaf area index (LAI), trees with high LAI are usually solid and transmit little light while as orchard with low LAI are likely to transmit through the tree. Light transmission through the tree is related to the density of the canopy. More light is transmitted with few leaves on the tree than through a smallar tree size with high canopy density.

Horizontal pulling/ training the branches

Pulling the branches in to the horizontal position, changes the hormonal make up of the tree or decreasing the Gibberellic Acid production sites by repeatedly pruning the young branches or by applying GA synthesis inhibitors. Pulling down the scaffold to the horizontal position also increases the ethylene content (Robitaille, 1975, Robitaille and Leopold, 1974) which usually stimulates flower bud formation.

Short internode

Variable internodes are found in different tree species. The internode length in apple for spur type cultivars is 22-25 mm while in traditional varieties it is 30 mm. As per report apple tree with internode length 15 mm are productive and those with internode length shorter than 15 mm for because of any reason are not productive.

Harmones in tree architecture development

Indole acetic acid produced in the vigours upper shoot tip and younger leaves moves downward in the phloem and modified the growth of the shoots tips and branches below the point of origin of the auxin, relatively large amount of auxin produced and transported downward along with other hormones affect the growth. Concentration of auxin hormones on the shoot tips lead the lateral bud growth inhibited, crotch angles between main stem and side branches are increased. Crotch angles are quantitatively related to the amount of auxin supplied from shoots directly above (Verner, 1955). The growth characters related natural auxin can be altered by pruning, girdling or application of hormones.

Branch angle

Some tree species have lateral branches with 90 degree angle or horizontal branching, which are important characteristics from bearing and tree growth point of view. For promoting wide angle branches, growth regulating chemicals i.e. benzyladenine can also be used. Wide angle branches have less amount of bud dormancy (apical dominance) while narrow crotch angle branches have relatively more Indol acetic acid content which is contrary to the wide angled trees.

Basitonic tree characters

Some of the tree species have strong apical dominance resulting bud break at the bottom of the branches. In basitonic branching habit, establishes a scaffolds system very close to the base of the tree resulting the smaller tree stature. Apple variety Mc Intosh has strong basitonic characters. Mostly basitonic trees develop short internode and usually have many strong internode.

Vigor management of fruit trees by horticultural techniques

Dwarfing rootstock- Dwarfing rootstock regulate the vigour of the scion. In fruits mainly apple, cherry, pear, mango, guava, citrus fruits growth regulating rootstocks are available which are being used in almost every parts of the world for controlling the scion vigour for ultimate canopy growth control.

Branch angle manipulation for vigor control- It is well established fact that vigour is higher at the top of the tree and much low at the close to the base of the canopy. It has been observed that branch bending at the bottom will impose more dwarfing effect than bending branch at the top of the tree. For attaining the uniform vigor in tree the lower branches should have a more upright angles and the upper branches in more horizontal angle. Bending changes the hormonal make up of the branches which creates a desirable effect but also causes some problems. Apical dominance is terminated with the branch pulling, the bud located on the basal portions of the shoot will break and produce strong upright shoots. Pulling the branches to the horizontal positions has been utilized to develop many canopy designs which are as follows.

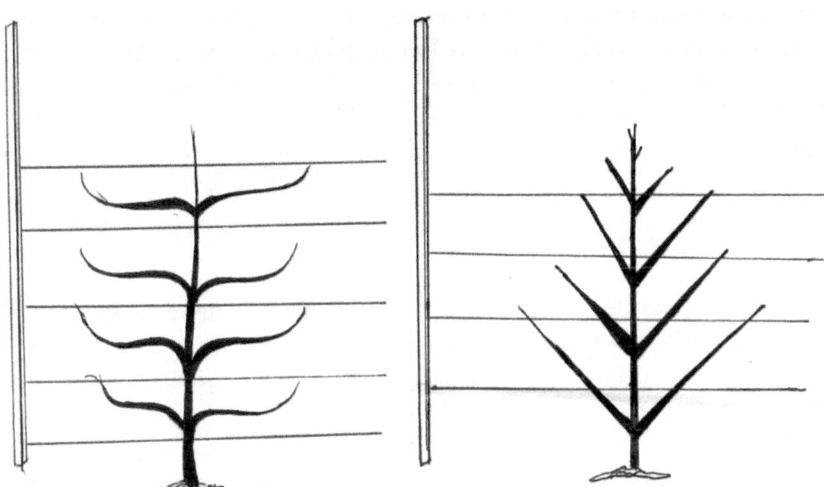

Fig. 1: Haag trellis **Fig. 2:** Palmette

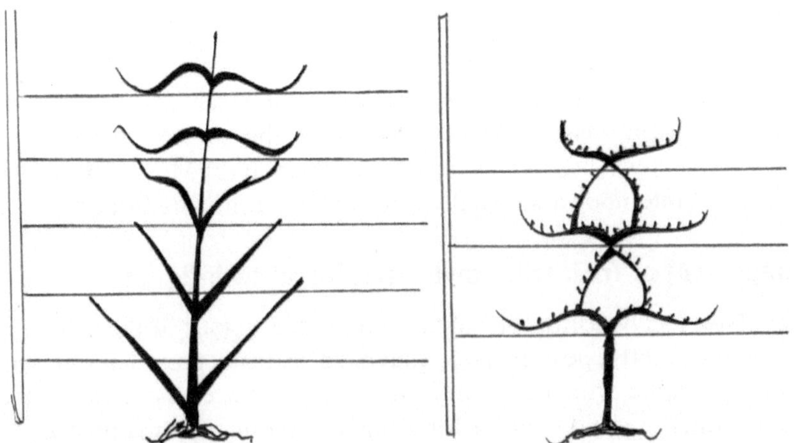

Fig. 3: Modified Palmette **Fig. 4:** Vincent Trellis

Ref. Different tree designs utilizing branch angle

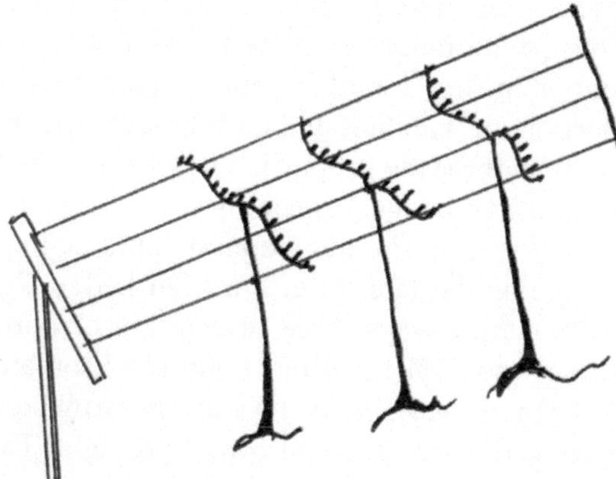

Fig. 5: Lincals trellis

Light interception

Light interception is the amount of available light intercepted by the tree canopy and not striking on the orchard floor. It is the function of plant density, branch and tree orientation, canopy shape, canopy size and leaf area index (Leaf area per/m^2 ground area) within the canopy. The dry matter of many other crops appear to be directly proportional to the interception of radiant energy (Duncan *et al.*, 1973; Loomis and Gerakis, 1975; Monteith, 1977; Gallagher and Biscoe, 1978). In annual crop the greatest loss of the light interception occurs at the beginning as well as at the end of the growth. Such crops frequently intercept virtually all the available light at full canopy (Sceicz, 1975) but may be slow to attain this because of delayed leaf emergence and slow leaf growth in spring (Watson, 1952; Sibma, 1977), during autumn, senescence of leaves may reduce interception while conditions are still suitable for growth. On the other hand-orchard crops tends to attain their maximum leaf area by mid- summer but usually intercept not more than 65-70% of the available light at the full. The amount of light interception by the tree is affected by canopy size, leaf surface area within the canopy and row orientation. Wertheim *et al.* (1986) reported that light interception by tree was linearly proportional to tree density. Light interception ranged from 57 to 81% of available light varied with orchard site due to tree size and cultivar differences and was reduced by summer pruning. Palmer and Jackson (1974) reported that production in a young high- density orchard increased linearly with increasing light interception between 20% and 60%. During initial stage of orchard establishment in various training systems, early production was correlated to the total light interception (Barritt *et al.*, 1991). Orchard systems such as a full-field or meadow orchard that have no alley-ways may intercept substantially more light than the conventional system.

Canopy shape affect the light interaction, it is proportional to the area of ground covered by the tree and the hedge height in relation to clear alley-way. Interception is more dependent on the horizontal orientation than vertical of the tree. The maximum proportion of incident light of branch can be intercepted by a hedge row is a function of (a) the proportion of ground which is directly covered and (b) the height of the hedge, especially of its "outside" edges in relation to the clear alley-way width. The interception could be high as the clear alley-ways width as wide as the later (e.g. 4m high hedge, 2m thick and separated by 2m wide clear alley-ways. Tree orientation of north to south intercept about 89% of direct and 87% of diffuse light (Jackson and Palmer, 1972). Wertheim *et al.* (1986) reported single and multiple-row beds of trees trained to two systems, taller and wider tree intercept more light. The multiple row bed (two to four rows), intercepts 10% more light than single row (Jackson, 1980).

Light interception measurement

Light interception varies during a season, leaf surface area development, solar angle and with distance from the centre of the tree or row and time of the day. The light intercepted in an orchard can effectively be measured to determine the intensity of transmitted radiation over a representative part of the orchard floor, in relation to available radiation over the canopy. In continuous hedgerow orchard, when tree develops uniform height and thickness a single evenly-spaced line across row centre to alley centre may be a representative sample for measurement of light interception. In orchard canopy (discontinuous type), the light sensors should placed in a way to cover the ground area across the alley between the centre of the trunks of adjacent pair of trees. Total light falling on the canopies are not utilized for photosynthesis purpose only appropriate wave lengths are productive. To measure the accurate light interception the sensor is placed in such a way to receive good angular response of light. It is worth mentioning here that the sensor has an appropriate wave length response. The photo synthetically active radiation (PAR) is 400 to 700 n m wavelength.

Photosyntheticlly active radiation (PAR) can be measured by LI-COR line Quantum Sensor (LI-1991SB) held horizontally in both a north-south and east-west direction with the centre of the sensor positioned adjacent to the trunk. The north-south and east-west data are summed to work out average and compared to the above canopy readings to calculate percentage of available PAR as measure of light distribution for three canopy positions i.e. top, middle and bottom of each central leader in tree (Barritt, *et al.*, 1991).

Canopy management principles are guiding force of efficient canopy development exploitation canopy of cultivar genetic potential to its fullest level thus harnessing natural resources.

Pruning influence light interception

Pruning usually reduces total light interception because of reduced growth and ultimately canopy size. Summer pruning has been shown to improve light penetration within the tree canopies (Marini and Barden, 1987). However, Wertheim *et al.* (1986) reported that summer pruning in combination with dormant pruning reduced light interception by 15%, one of the benefits of summer pruning is an increase in photosynthetic photon flux density (PPFD) within the tree canopy. Porpiglia and Barden (1981) reported that after 2-3 leaves shoot pruning in August, PPFD was increased both at periphery and within the canopy. However, following summer pruning, PPFD was lower in summer pruned than control trees which were less severely headed back in dormant season. Morgon *et al.* (1984) studied the effect of summer pruning and found increased PPFD penetration in summer pruned trees. Taylor and Ferree (1984) also found that summer pruned trees were more open in September, October than unpruned. Summer pruning increased light levels within the tree canopy during the season of pruning but may decrease levels during the following years. The summer pruning resulted in less canopy shading but poses certain other problems. The major advantage of summer pruning may be that pruning can be spread out during the year whenever labour is available. Summer pruning may have a place in training of fruit trees and may also be used to improve fruit color for poor coloring cultivars grown in climate where color development is less than ideal. Further, summer pruning must be used with due care to avoid reduced fruit size or soluble solids. The major benefit of summer pruning is to suppress the tree vigor and up to some extent opening of small hole for penetration of sunlight, which discouraged the development of micro-climate congenial for disease and insect-pest growth.

Physiology of pruning

Pruning is done to achieve one or more objectives like enhanced fruit quality, develop balance between vegetative and reproductive growth, making tree canopy under control, inducing dwarfness. We remove wood, buds and cambial tissue. We remove carbohydrates and nitrogen amount by pruning of wood that could have served as the reserved materials for next year growth. Removal of buds in pruning led to reduction of leaf area, and fruit and removal of cambial tissue, loss of secondary growth. Shoot growth in pruned trees significantly increased as compared to un-pruned trees, however, trunk and root growth are greatly decreased (Forshey *et al.,* 1992). Usually shoot growth is stimulated by pruning, particularly heading back of stem and removal of terminal shoot portions and lateral buds etc. Heading back cuts altered hormonal balance among the remaining buds on the shoot, while as in thinning cuts where entire shoot is removed, did not alter hormonal status. Heading back and thinning out results removal of terminal meristem which are the seat for abundant auxin contents and undisturbed root tissue continue to produce cytokinin, the harmonal balance inclined towards cytokinin which stimulates cell division ultimately shoots growth. The apical portions of the fast growing shoot apex, starts producing auxin and gibberellin in the large quantities which enhanced shoot growth. Heading back of branches, removes terminal meristem and inter-related apical dominance. Removal of apical branch portion leads to proliferation of lateral buds, resulted excessive shoot growth and bushy type canopy development. According to Faust (1989) distal buds on a shoot have greater growth potential than basal buds. Similarly the buds which are located higher on the canopy have strong, growth potential than lower buds. Changing the orientation of branches from normal vertical growing shoots to a more horizontal position which ultimately reduced the rate of bud growth.

Source-sink relationship

Photosynthesizing organ in plant is known as source, mainly mature level, produce photosynthates, especially carbohydrates. Growers tries to manipulate the source-sink relationship to achieve adequate fruit production and quality (Gil, 2006). According to Park (2011), balance between vegetative and reproductive growth of a tree is of great importance for growth and fruit production. Training and pruning alters the balance between vegetative and reproductive growth by allocating the carbohydrate, water and growth regulators (Myers, 2003). Heavy pruning diminishes leaf area, whole tree photosynthesis and translocation of photosynthates to fruit and roots, increasing the root, shoot ratio (Casierrt –Posada and Fischer 2012) and favoring growth. During the reproductive phase, pruning, helped to improve fruit load, regulate the physiological balance (vegetative and reproductive), ensures a harmonious and rational distribution of high quality fruit (Arjona and Santinoni, 2007).

Carbon dioxide assimilation

There was decreased in internal CO_2 in non flowering branch as compared to flowering. It also indicated that there was faster rate of assimilation of CO_2 in leaves of pruned branch as compared to un-pruned (Singh and Singh, 2007). At 50-100 ppm of CO2 level, no photosynthesis took place. The rate of photosynthesis was gradually increased with increasing concentration of CO_2 and maximum photosynthesis (15 m mol M^{-2} S^{-1}) was obtained in 450ppm and 350ppm concentration of CO_2 in pruned trees of Allahabad Safeda and 'Sardar' guava respectively. In case of high concentration of $CO_{2,}$ inhibition in photosynthesis rate was recorded; Photosythesis started declining after 450 ppm concentration of CO_2. Increased in the photosynthesis rate might be due to increased in the efficiency of Rubisco under elevated CO_2 or reduction in photorespiration rate (Acock and Allen 1985).

Light penetration and distribution within the canopy

Light intensity decreased within the tree canopy as the outer portion shades the inner canopy. In central leader training system, light sharply decreased inside the first 0.5 to 1.0 m of canopy from canopy edge. In the bottom of 4 m tall trees, light was reduced to 8 to 10% of full sunlight (Barritt *et al*., 1976). Jackson and Beakbane (1970) measured the light intensity on the leaves of Cox's Orange Pippen apple at a range of positions within a large tree, using sensors which gives an equal response to light at all angles above the horizontal and compared this to the thickness of leaves and thickness of their palisade layers. Both were positively and linearly correlated with light intensity. Shading reduces specific leaf weight, Gabrielson (1984) reported that shaded leaves possess maximum energy, however the yield considerably less than that of exposed leaves. Barden (1978) reported linear relationship between specific leaf weight and rate of the net photosynthesis under standard conditions. Leaves growing in full sun take advantage of maximum photosynthesis. However, Arden (1977) found that shaded leaves have similar amount of photosynthesis to sun exposed leaves at low irradiance, which indicates that leaves even in heavily-shaded position contribute positively.

Light transmittance through the canopy decreases with increasing LAI (Heinicke, 1963) and follows Beer's Law (Jackson, 1980). Light transmittance is a logarithmic function of the leaf area index, of the distance through the canopy that the light penetration depend on incoming light intensity. Light intensity decreases from the top of the tree canopy to the interior bottom of the canopy. The light environment within the canopy is not uniform. The inner portion of the canopy may receive high light intensities at the times due to sun flecks penetrating the canopy or change in solar position as the day advances. The portion of canopy may receive high light for a brief time and predominantly low light for remainder of the photo period. Similarly, proportion of a direct vs. diffused light changes with diurnal changes in the light intensity (Lakso, 1975). Full sunlight gives an illuminance of about 10000 Im/ft^2 (foot-candle), i.e. 107,600 Im/m^2 (lux) and an irradiance of about 1.3 cal/cm^2/minute nm wave end, the photosynthetic active radiation (PAR) is from 380 to 720 nm. Net photosynthesis is considered to be saturated at 90 to 99% of maximum pn (net-photosynthesis). Saturation take place at 700 to 1100 u mol. S^{-1}.m-^{2} or 30 to 40% of full sun. The saturation point and pn rate vary with environment in which leaf develops and type of leaf. Leaves developed lower light levels have lower net photosynthesis and lower saturation points leaves, where photosynthesis balance respiration, occurs at 2.4 to 7 w/m^2 of 400 to 700 nm irradiation (less than 2% of full sunlight), photosynthesis of 80% could be attained between 10 and 40% of full sunlight. Leaves in full sun reached their maximum photosynthesis rate at about 75% of full sunlight and, leaves in shade attained the light saturated photosynthesis level from 44 to 75% of full sunlight. Lakso and Seeley (1978) reported that single leaves attained about 80% of their maximum photosynthesis rate at around 25% of full sunlight. Further, Lakso and Seeley emphasized that whole tree photosynthesis to be about 60% of the maximum at 25% of full sunlight.

Objective of canopy management is to maximize the light penetration and distribution inside tree canopy. Total light interception of an orchard can be raised by increasing the density of the canopy. Recently, there has been a change from standard trees planted at low densities to more efficient orchard systems on account of various cultural management problems and low levels of productivity. In horticultural advanced countries orchard systems have been developed which minimize the cultural operation and improve profitability.

The current research on apple canopy management depicts that the widespread adoption of slender spindle and north Holland spindles as tree forms with multi-row beds in combination with vigor-controlling practices like use of growth retardants, training, pruning systems and dwarfing rootstock.

Effect of tree architecture: Cumulative production to the sixth year for central leader was 41% of the field for the slender spindle in 'Golden Delicious' (Ferree, 1980). The development of new tree architecture has resulted in maximum light interception and distribution'. Since light interception is a function of tree planting density. training and pruning systems for each cultivar-rootstocks combination. Tree density must be established to allow trees to rapidly fill allotted space after planting. During orchard establishment process and attain 70-80% light interception. Two approaches are in practice to improve the light distribution in tree canopy, one is to use relatively natural tree forms that allow light penetration through the canopy by providing many small openings in the foliage such as in multiple leader, central leader, vertical axis, or slender spindle forms. This approach requires a high degree of horticultural skill to manage the canopy of the tree and the second approach is to provide fewer, large and permanent openings for light penetration into canopies restricted into geometric forms. The thin restricted planes of foliage such as narrow hedge rows, tree walls, and A, V or T forms. For development of such canopy model requires high class horticultural skill, expensive support structures and intensive labor requirement to maintain and place the branches in specific locations. Thin vertical canopies receive light exposure from both sides of the canopy, hence, having good light distribution within the canopy. However, palmette hedge row had better light penetration into canopy than slender spindle or pyramid hedgerow trees. Trellis had greater crop density and higher efficiency than the other training system. On the basis of experience gained from various advanced training systems, of the trellis canopies are kept thin, they produce excellent fruit quality as the width of the canopies increased, the centre of the canopy produce poor quality fruits. The vertical staking of fruiting branches with palmette trellis hedge rows leads to more growth on the top branches than on the bottom branches due to better light exposure. If this is not carefully managed the top growing branches shades the lower ones, resulting in poor quality fruits. Inadequate light distribution/penetration within tree shade can limit yield by reducing fruit number, size and quality. An optimum light level of near 70% full sun may be necessary to maximize the most sensitive quality criterion and fruit color. Canopy height and width affect economic yield of tree, which could be managed according to the ambient maximum irradiance and latitude of the orchard location.

Canopy development of orchard

Development of canopy for newly established orchard and full leaf area index depend on the tree vigor, spacing and management. If trees are planted at spacing appropriate to their potential size, the orchard will attain its maximum leaf area index and their full growth. In inter cropping of low growing fruit crops between the main crops the filler need to be uprooted as the tree began to compete for light with each other and become so crowded that pose problem in management. As the technical know-how about the canopy concern are increasing the orchardists deliberately encouraged the lateral growth in fruit plant from nursery stage and planted in very close spacing to utilize the allotted space at earliest and utilize horticultural technique to control vigor. It is a well established fact that the larger trees are planted in relation to their ultimate size, the more rapidly LAI are obtained. The objective of attaining quick canopy size can by fulfilled by using vigours rootstock which required high skilled horticultural technique for controlling the growth of the tree. Total light interception in north-south row orientation did not vary after complete canopy formation (Jackson and Palmer, 1972). While east-west rows were inefficient at midseason light interception but due to solar angle, tall east-west hedge-rows intercepted more solar radiation than north-south hedgerows at higher latitude during early and late summer.

3

Canopy Manipulation for Optimum Utilization of Light

Canopy manipulation in fruit crops has a seasonal and a life-time development pattern and have lifelong advantage and disadvantage. The unmanaged canopy of fruit trees poses many difficulties i.e. difficulty in spray, pruning and hand harvesting, poor light penetration and interception throughout the canopy. Low early light interception, leaf area index and fraction of land covered by canopy, leading to delayed and poor cropping. These disadvantages have resulted in widespread effort of manipulation of canopy model of the perennial fruit trees by increasing tree density, tree size reduction and use of other horticultural techniques to increase the light interception and distribution.

Now a days canopy manipulation in tropical, subtropical and temperate fruit deserve the utmost important. Several horticultural approaches could be utilized to control the tree size. Canopy manipulation, includes the use of rootstocks and scion for vigor control, pruning and training, growth retardants, suitable nutrient and water management practices and location of orchard site. The horticultural techniques is practiced to modify the tree canopy.

Light distribution in different canopy forms

On the basis of canopy profile exposure to sunlight it can be divided into three categories; i) well-exposed zone with light level more than 50% of full sunlight' ii) marginally exposed zone, light level 30-50% and iii) well-exposed zone with light level less than 30% (Fig.13). Increase in tree height increased light interception particularly for tree of triangular section.

The interior micro-climate of central leader training systems rapidly declines the light level in inner portion of the canopy due to seasonal changes in growth

Fig. 6A: Light distribution in a large central leader tree (Looney, 1968)

Pruning

Pruning is the removal of plant parts to achieve a desirable architecture of the tree and to lighten the foliage density by removing the unproductive branch of tree. It is process that may influence the physiology of the tree for several years. Trees portion or some parts are cuts with an objective of altering the physiology of the tree. Pruning is done in young plants to develop functional frame work while in older tree cutting away to expose tree into more solar light in the inner parts for increased aeration in the canopy. Pruning is a procedure to dwarf the tree and has been widely used in different fruit crops.

Type of Pruning

Different types of the cuts are applied on a tree, shoot shortening or thinning out are two most popular pruning techniques i.e. shoot shortening involved tipping, heading back, stubbing, representing the severity of the cuts. In summer pruning, shoots are normally pruned, pinched, tipped or headed back while in winter they are headed or stubbed. Removal of entire branch from the base is called thinning out.

Pruning performed on the dormant or active growing shoots remove the apical dominance. The removal of unexpanded leaves enough to induce bud growth (Mika, 1971), while removal of expanded leaf did not stimulate the axillary bud growth. Heading back of dormant shoot changes the growth pattern of the branch. On headed back shoot, the first bud produces upright growth and the lower buds develop in to side branch with wider crotch angle. The angle of the laterals at the base of the tree trunk are allowed to grow as narrow crotch, while successive tiers grows naturally wider and at the top of the tree the angle is almost horizontal. Heading cuts also influences photosynthesis by forming new sink in form of several growing points.

Pruning severity

With increasing in pruning severity, new shoot growth increased while as the circumference of the trunk growth decreased. Increased in pruning severity the root growth decrease. Tree response to the pruning is also influenced by the size and type of cut, many small cuts are better than one severe cuts, as many small cuts stimulate growth more than a few large cuts in balanced form (Mika 1982). Heading back cuts stimulate more growth than complete thinning out (Mika, 1982). Both dormant and summer pruning increased light interception in the canopy which influenced the leaf structure and photosynthesis. In temperate climate investigations on the branching process of fruit trees have been focused on one year old shoots, because apical dominance and bud dormancy occurs at the level of organization. Apical dominance is the control of apical portion of the shoot on axillary bud during growing season. Apical dominance involved different correlative mechanism mediated by auxin and cytokinins (Cline, 2000; Sussex and Kerk, 2001).

Mechanical pruning

The un-pruned tree, which has well distributed vigor, fruiting shoots, quickly starts production. However, in overcrowded canopy architecture, fruit size and quality are reduced after some times. Orchardist prune the tree to make the balanced growth between vegetative and reproductive growth to produce annual quality fruits. Two types of pruning systems are in practice, dormant pruning and summer pruning. Dormant pruning of fruit tree is designed to ensure light penetration. Mika and Antoszewski (1972) reported the effect of pruning to ensure a loose crown for light penetration. Total light radiation within the tree canopy increased with increased distance between framework of branches (Rud *et al.*, 1976). Moderate thinning and opening of the crowns of trees planted at a distance of 8 x 3 m and 8 x 2 m increased light penetration so that photosynthesis within the crown increased by 4.5 times compared to control (Kudryavests and Trusov, 1975).

The mechanical pruning was tried in many countries but results were not found encouraging as it gives excessive number of well-illuminated one-year-old shoots and sets too many fruits which involve greater cost of thinning out. Possibly the effects of both absolute light levels and the ability to crop on one-year-old-wood are involved. The greatest effect of pruning on light interception and penetration in orchards in recent years followed the concept that a relatively shallow depth (1-1.5 m) of fruiting canopy was capable of intercepting most of the available light and only upper and outermost parts of large round canopy trees received sufficient light to produce good quality fruits and fruiting buds. The conventional large round canopy tree has been permanently lowered successfully over a period of two years, without any loss in yield to a convenient height for performing the cultural practices (i.e. pruning and spray operation) with the use of NAA paint on the pruning cuts to avoid vigorous re-growth and shading. Approximately 75% of orchards suitable for re-framing in this way have been so treated and fruit size and color improved.

Summer pruning

Summer pruning by removing the vigorous growing shoots increased light intensity in cropping zone and color intensity. Late summer pruning also reduced the growth due to reduction in photosynthetic capacity and ultimately carbohydrate reserve by reducing the LAI and the spread of the canopy. The orchard of dwarf trees generally do not intercept as much light as large tree orchard can, but their yield can be high. Mckenzie and Rae (1978) reported that maximum yield from summer pruned dwarf-pyramid tree (180MT/ha) in 'Golden Delicious' of apple. Summer pruning may have a place in training of young fruit trees; however, it must be used with caution to avoid reduced fruit size and soluble solids.

When to pruning

Pruning timing depend on the pruning objectives, at the time of newly plantation pruning is carried out to develop well structured frame work. Any damaged, overlapping, diseased, dried branches must be pruned. Mostly pruning is carried out when plant is in rest phase or just before on set of new growth. During dormancy (devoid of green foliage) helps in selection of branches. Dormant pruning is done to invigorate the plant, while as summer pruning reduce the growth and excessive growth etc. Too early pruning in winter must be avoided as it leads to winter injury. Older trees in orchard must prune later, while newly planted young tree early in winter season. Pruning induces flower bud initiation. The plant which experience late flowering must be prune first than early blooming trees. Pruning during summer or active growth phase must be limited to removal of thinning out cuts to reduce excessive vigour and upright growth.

Tree architecture development

Tree architecture development (training) is the improvement in integration of different components of the orchard during tree life i.e. quality, density, rootstock, cultivars, pruning, thinning out practices. Tree architecture is thus required to manipulation of plant with aim to obtain maximum quality fruiting. High yield performance of fruit crop results from combination of various components viz- rootstock, training system, cultivar, density, tree arrangement, pruning, thinning out. Training of fruit tree is important component of production system it is carried for improving for light trapping for high biomass production, increasing the canopy porosity for light (Lakso, 1994) improve in light distribution between fruiting structures (Lakso and Corelli-Grappadelli, 1992); Winsche and Lakso, 2000) and partitioning of the biomass to the fruiting point. Light interception by the individual tree depend on canopy form. In nature different canopy shape develops viz. Pyramid, conical, oval, spreading, vase shape weeping as well as round shape canopy.

Physiology of tree architecture development

The objective of training or tree architecture of the newly plated saplings is to develop strong framework with minimum limb breakage and maximum light interception. High light perpetration into the interior portion of the canopy results good flowering and fruiting. Open canopy also allow good air drainage and reduce microclimate development congenial for pest and disease growth.

The scaffold branches should be encouraged at 60-70^0 angle from the trunk, which lead to strong attachment on the point of joining. Main scaffold branches must keep at 20 cm interval or 60 cm apart on vertically along with trunk spirally, almost 4-6 scaffolds develop balanced canopy.

Training methods have been developed at the tree scale to manipulate vegetative growth, flowering and fruiting aspects. Bending or tying of branch is employed with objective to contain the tree with in the allotted space. Thinning out and heading back cuts are made to control the tree growth and shape and is particularly used for specific type of training procedure for effective utilization of natural resources, training and management systems at orchard scale and training method (pruning and bending) at tree scale. Training of the perennial trees to the open centre has always been an age old practice to harness the advantage of light and ventilation. The canopy design and shape greatly influence light interception. Light interception is proportional to the ground covered by the tree and hedge height. Interception depend more on the horizontal than vertical components of the tree size. Tree training to multiple leader, central leader, vertical axis or slender spindle, forms many small openings. However, thin restricted canopy like narrow hedge rows, tree walls, A, V, or T forms few large permanent openings for light penetration into canopies. Modern training systems are more advantageous and efficient from light penetration and interception point of view but it involves severe geometric restriction of the canopy, strong support structure, and skilled manpower to maintain the branches in the specific directions.

Central leader tree architecture

This tree canopy model has been widely used in areas which are more prone to heavy snowfall, leading to breakage of branches. Central leader system forms a pyramid-shaped canopy with tiers of branches spaced along the trunk. The widest part of the tree is at the bottom tier (Fig). Years of experience has led growers to increase distance between the bottom and second scaffold branches to at least 1.0 m to leave gaps in the canopy to increase penetration of light to the bottom scaffold. Summer pruning is done in central leader canopy architecture to increase light interception in the interior parts of the canopy Fig 7).

Fig. 7. Central leader tree architecture

Open center leader architecture

This canopy form is suitable to the region, where sunlight is limiting factors. Open canopy allows good air circulations thus reduce disease and insect growth. Since most of the canopy portions are well illuminated hence, good yield of quality fruits are harvested. Such canopy permits maximum penetration of any chemical spray (pesticides and plant growth regulators). One year old, well feathered saplings are planted in the field, after planting main trunk is removed at 50-75 cm from ground level and allow 4-6 primary branches in all the directions at 10-15 cm interval. Secondary branches arise from the primary scaffolds and in 2nd year main scaffold branch is removed. During pruning in dormancy season misplaced shoots arising must removed, complete canopy develops in 2-3 years (Fig 8).

Fig. 8. Open center leader architecture

Modified center leader type architecture

This type of tree architecture is most commonly used in India due to combined open center and center leader system of training. It is similar to center leader type advantage of architecture except the main scaffold are modified or removed after 3-4 years, laterals are allowed on the main scaffolds in spiral fashion (Fig 9).

Fig. 9. Modified central leader

Round-headed tree architecture

In spherical or hemispherical tree canopy, only a limited outer zone of such trees receives enough light to produce quality fruits. Heinicke (1963, 1964 and 1966) reported that large-headed apple tree fell 42% full sun at 2.0 m depth from the top of the tree, the fruit growing below this level (below 2.0 m) receive poor light intensity leading to poor color development. In general best color in apple develop with light exposure of more than 70% of full sun, adequate color from 40-70%, and inadequate color with less than 40% full sun. The fruits receiving less than 50% of full sun were small and the soluble solid content in fruits were higher where they were well illuminated to light. Heinicke (1964) found as the tree size decreased the heavily-shaded area within the canopy decreased. Dwarf trees with some overlap of canopies had 1/3 more leaf area/ha that received more than 30% of full sunlight than the standard trees, ultimately increased photosynthesis and good yield and quality. The large and small tree had the same dry matter yield per unit area despite lower LAI on dwarf tree (Forshey and Mc Kee, 1970). Such trees have more efficient leaf surface, thus produce 80% more fruit per unit area than big tree. The decrease in efficiency of larger trees is due to the result of greater internal shading.

Palmette leader architecture

It is the modification of the central leader system, trees are either fan-shaped or have a vertical central leader or leaders from which scaffold arms may arise to right or left and in both directions. The tree, thus has two dimensional aspects namely height and width, but not canopy thickness or depth. Palmette leader is designed to improve the light distribution in the tree canopy. A large gap is created by removing upper east and west growing branches and results in a flat north-south oriented palmette top. The primary advantage of the palmette leader canopy is large gap in the east and west sides of the tree guarantee good light exposure to all parts of the tree and good light penetration in the center of the tree throughout the growing season (Fig.10).

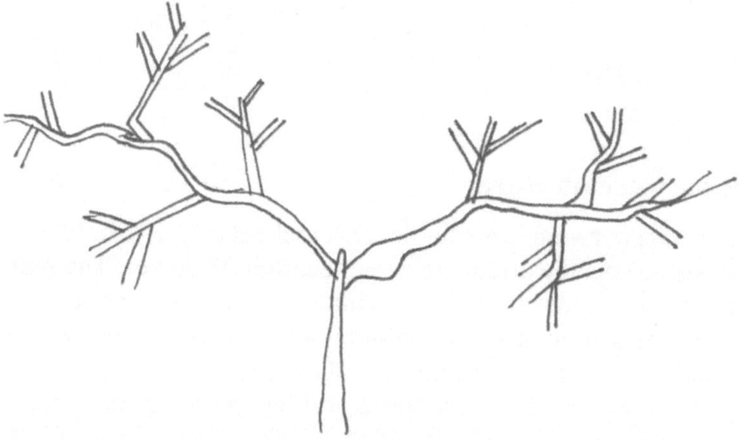

Fig. 10. Palmette leader tree architecture

Dwarf pyramid architecture

A low- headed compact, central leader tree, with lowest branch at 30 to 35 cm from the ground and with successive branches radiating at intervals along the main leader, gradually diminishing in length from bottom to top. It is typically suitable for apple, but pear, peach, cherry and plum can also be trained. The central leader is not allowed to attain height of more than 2m, through out life of the tree. In perfect pyramid canopy, the lower most branches should attain a length of 1m and the middle branches 60-80cm. while as the upper most branches about 45cm length. Summer pruning is essential to maintain light penetration in the canopy (Fig. 11)

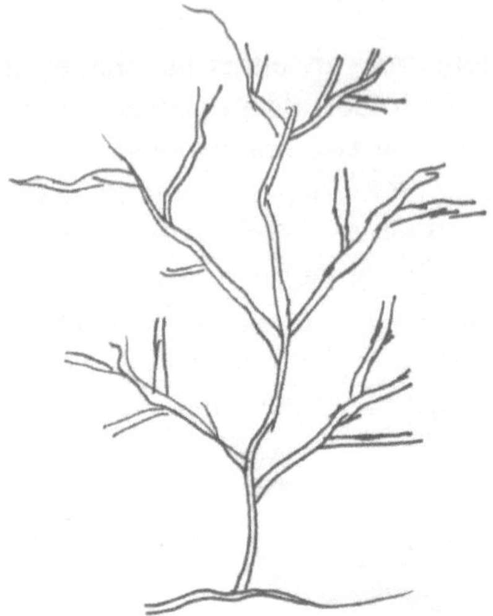

Fig. 11. Dwarf pyramid architecture

Slender spindle bush architecture

A modification of dwarf pyramid is a spindle bush architecture, it differs from the former that it has no specific arrangements of scaffold branches. The most important feature of the spindle bush is its tying down the lateral shoots to horizontal position. Trees trained in this system starts bearing fruits within 2-3 years after planting. Well feathered nursery plants are planted in the field, and headed back at 60 cm, with lowest shoot at 30cm above ground level. The central leader is encouraged and maintained. Single stakes 1.5 to 2m long above ground level be erected near tree to make the crotch angle of 30^0. The central leader is removed when tree attained 2-2.5m height (Fig.12)

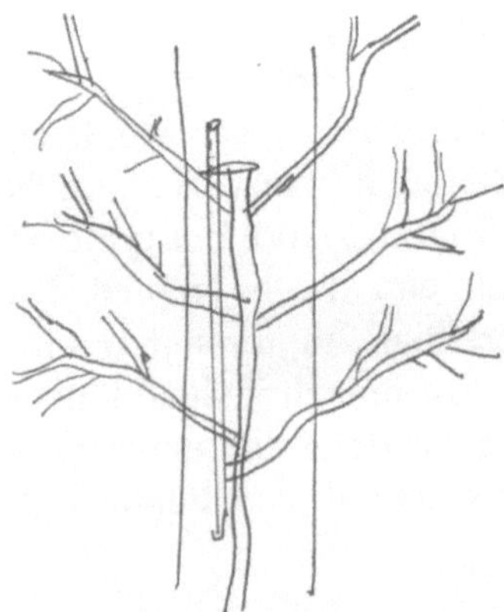

Fig. 12: Slender spindle bush canopy

Trellis Canopy

This system was developed by Irrigation Research Institute from Tatura, Australia, in 1973 consists of rows of 'V' shaped trees in north-south rows. Each tree has only two limbs growing east and west at an angle of 60^0 to horizontal position. In trellis system tying down of branches to horizontal position causes a gravemorphic response which leads to premature terminal bud formation and sometimes to improved flower initiation.

A high degree of ground cover combined with a shallow canopy can be developed by growing relatively vigorous trees in hedge-row system with branches trained in form of an open V or T. This canopy is suitable for mechanical harvesting but leads to excessive upright vegetative growth. The shallow canopy insure high light interception with uniform irradiation on large canopy surface. Trellis canopy results in precocious and profuse fruiting in many temperate and subtropical fruits (Fig. 13).

Fig. 13: Tatua trellis (Chalmers and Vanden Eude, 1975)

Solen system

The 'solen' a low domed form of tree architecture is particularly adopted to tip bearing apple cultivars tree like Granny smith. It is particularly suitable where climate and soil increase vigor (Lespinasse, 1989). It facilitates harvesting and pruning. In this system two scaffolds are promoted and tied the first two years after planting. On both sides of the arms, 12-15 fruiting branches are developed. The height of the tree is between 1.2 and 1.5m according to the cultivars. This system is mainly suitable for dwarfing rootstocks like M-9 (Fig. 14).

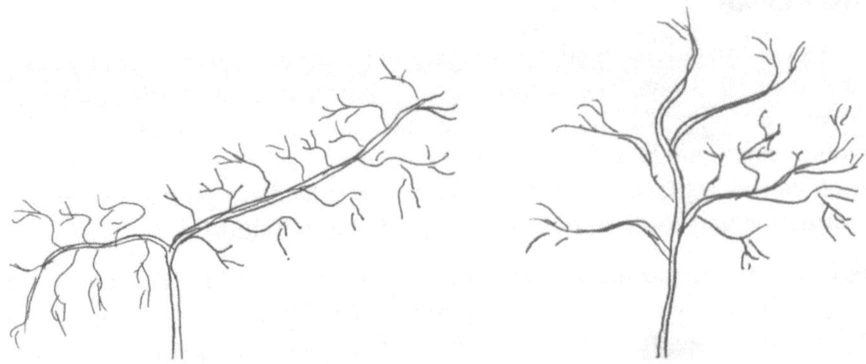

Fig. 14: The 'solen' a low domed form of tree architecture

Multi-rows and bed system

In order to achieve high light interception with even light distribution in the canopy, Palmer and Jackson (1978) developed a bed system for apple with a more even distribution of foliage over the orchard floor. In the tree bed system, trees spaced at 1.5 x 0.5m in 14 row bed separated by tractor alley-ways. It achieved ceiling levels of leaf area index (LAI) and light interception in discontinuous orchard canopy, however LAI/light interception relationship could be modified to some extent by leaf distribution pattern over the ground area. Palmer (1988) noted that bed system can intercept up to 80% of photosynthetic active radiation from the end of the June till October resulted yield of 78 ton/ha in 3rd year in apple. Difficulty with both multi-row and full field canopy system is the shading effect of the annual shoot growth, such a near-continuous-canopy system, forms a vegetative shading layer above the fruit bud and fruiting-producing zone. Summer pruning or chemical means of growth control may be profitable in multi-row or bed system of canopy. The genetic potential of spur and tip bearing apple cultivars can be harnessed in full field canopy in which fruits are produced on the upper un-shaded canopy.

Most revolutionary approach in multi-row and bed system planting is the meadow orcharding where 20,000 to 1,00000 trees/ha were raised. It is designed to produce fruits on two-years-old plants which are chemically regulated to produce a simple and smaller structural framework. Harvesting is by mowing the orchard with a combine harvester, hence known as meadow orcharding. The light values at the base of the canopy in the first year are well within the accepted limit for fruit bud formation (Bennett and Turner, 1978) but in the cropping year there is some cultivars, where fruit size and colors are reduced in the central of the plot due to lower light conditions (Child and Atkins, 1978; Bennett and Turner, 1978). The problem of meadow orcharding is the light use efficiency due to biennial crop. If it could be adopted as an annual system by tree training to two main shoots, one of these is headed back to form the above biennial cycle in one year and the other shoot is headed back in the following year, so that light interception and distribution efficiency be high.

High Density Planting System

High density planting systems is a type of multirows and bed system which accommodate higher number of plants per unit area as compared to the conventional planting density. High density planting envisages the planting of small trees densely, restricting their vegetative growth by use of dwarfing rootstock, growth retarding chemical, pruning and thereby diverting much of the plant energy to the economical parts. HDP system required to achieve twin requisites of productivity by maintaining a balance between vegetative growth and reproductive load without impairing the plant health. It may also be employed in different species and multistoried crops. In multistoried crop of light interception at different tiers by canopies of various species based on their light transmission characteristics and shade tolerances are exploited.

Principle of high density planting

The basic principal of HDP is to make the best use of vertical and horizontal space of the land and to harness maximum possible return per unit of input application. The multi species high density planting is practiced in the tropical and subtropical region of Asia and Africa, where farmers have been practicing inter-cropping to various degree and levels. Kerala and coastal region of Karnataka are typically practicing multi-storied cropping system to harness maximum solar incident energy by way of managing canopy model. Close planting of the tree and development of dense canopies may alter the micro-climate around the tree canopy. Temperature and light regime decreased while humidity increased which favors incidence of disease and pests are negative side of high density planting. Principle advantage of high density planting is better utilization of incident solar radiation and increased in bearing surface per unit land area.

Advantages of high density planting

1. The tree under HDP commence in fruiting early
2. Improved in fruit quality and yield
3. Efficient use of fertilizers, water, land, solar radiation, and other inputs.
4. Enables the mechanization of fruit production e.g. drip irrigation, fertigation, automation, tree training and pruning practices etc.

Disadvantages of HDP

1. High initial investment required and skilled manpower
2. HDP resulted in overcrowding, over shading and overlapping of branches.
3. High competition for root system, space, nutrient and water.
4. Build up of high humidity, lack of cross ventilation in orchard which promotes pest and disease growth.
5. Reduction in yield after 10-12 years.
6. Production of small size of poor quality fruits.

Influence of rootstock on canopy

In temperate fruit a wide range of vigor controlling rootstocks are available by which more number of plants per unit area can be accommodated, which intercept maximum solar energy resulted high yield. In tropical and subtropical fruits also number of dwarf and semi-dwarf rootstocks are being used to regulate the canopy size, shape and thickness. Rootstocks have been extensively employed as a means to regulate canopy vigor. It influenced size and shape of canopy of scion grafted on them and maintains shallow canopy or thin hedgerow which can be adequately illuminated.

Grafting of scion on different rootstocks used to increase orchard density and tree productivity (Fallahi *et al*, 2002). Dwarf rootstock reduce the overall tree volume and encourage precocity (Lockard and Schneider, 1981; Larsen *et al.*, 1992; Barritt *et al.*, 1995. Reduced scion growth on dwarfing rootstocks as well as grafted plants, resulted plant with small canopy management is easier, and tree occupies its allotted space quickly. Small tree needs less pruning, and for foliar spray.

Dwarfing rootstock control scion growth

Dwarfing rootstock have tendency to control the scion growth. It is said that larger tree have a larger root volume. According to Beakbane and Thompson (1947), Simons, (1987), different rootstock have different vascular anatomy. In dwarfing rootstock xylem anatomy is such that it reduces the quantity of water movement to scion, resulting scion dwarf. Dwarfing rootstock cause reduction of aerial growth due to delay of leaf emergence rate or to a shorter period of growth. The rootstock reduces the internodal length (Seleznyove *et. al.*, 2003; Wield *et al.*, 2003. The dwarfing rootstocks have lower xylem and phloem ratio than the root systems which don't have dwarfing characters. The dwarfing rootstock creates resistance in water movement at union portion (graft union) than non dwarfing rootstock, which results dwarfing effect on scion varieties.

Yield efficiency

Yield efficiency is a production factor which is controlled by scion and rootstock combinations, Scion growth get reduced on rootstock-scion combination than non grafted plants as dry matter produced, more partitioned to the fruit production and less to vegetative growth. Yield efficency is most reliable yield estimation. It is estimated by using yield kg/tree divided by trunk cross sectional area (TCSA), or yield kg/tree divided by per m^2 canopy area.

Growth regulators

Growth-controlling chemicals have been used successfully to retard the shoot vigor and canopy volume. Growth retardants have been employed in apple, peach, plum, citrus, mango, pear etc. chemicals like paclobutrazol, promalin, CCC, MH, AMO 1618, XE1019 are in practice to regulate the canopy size and shape. The plant growth regulator, prohexaiodone calcium experimentally known as BAS-125 (A pagee), developed in USA by BASF Crop has been found to be a promising chemical for controlling the vegetative growth of apple (Unrath, 1999).

Orientation of row and branch

Light interception and penetration into the canopy depend on the canopy type and distance between trees. The part of the canopy light interception depends on the orientation of the row in the orchard. If the rows are oriented to the north-south direction, both row side will receive the same light hours, while row oriented to the east-west and the north side of the trees receives 2/3rd fewer hours of direct radiation than the south side (Gyuro, 1974). The effect of planting density on light interception can be modified by the row orientation and rectangularity. North-south orientation and rectangular plantings intercept more light than east-west and square planting system. Shoot growth reduced proportionally to the increasing planting angles from vertical to horizontal position, hence minimum shoot extension growth was observed at 30^0 plantations in 'Royal Delicious' as observed by Sharma and Jindal (1992). Branch orientation from vertical to horizontal position, the growth reduction noted with increasing orientation (Hamzakhyal *et al.*, 1976). It is due to the lower auxin content in the top of horizontally growing apple shoots, than upright growing shoots (Kato and Ito, 1962). The supply of the nutrient to the apex is controlled by the auxin in top meristem (Luckwill, 1968). The tree orientation influenced the spur production (Sharma and Jindal, 1992). Tree plantation at 45^0 experienced maximum number of spur formation, than vertically grown trees. In horizontal branches, buds growing on the upper sides remained dormant or grew into spur, the variation in the lateral bud growth between upright and inclined branches results from the effect of gravity on the distribution of growth regulating substances within the stem (Sharma and Jindal, 1991), Several researchers have investigated the mechanism by which shoot re-orientation favors flower production (Ito *et al.*, 1999, 1972; Wareing, 1970 and Yang *et al.*, 2002). Floral development progresses more rapidly on horizontal than on vertical shoot (Ito *et al.*, 1999).

Nutrition

Nutrition has a significant influence on the growth of the shoot in tropical, subtropical and temperate fruits, trees. Even the fertility level of the soil and climatic conditions influenced the tree vigor. The nitrogen favored the growth and increased the canopy vigor and volume, potassium reduced them (Shikhamany, 2001). Nutrient application in non bearing trees resulted unwanted vegetative growth, it must be provided as per growth and bearing state of the tree. Nutrient must be applied rationally at recommended dose and recommended time. Excessive nutrient application resulted excessive. vegetative growth and canopy imbalance with more of vegetative growth and less flowering and fruiting.

4

Canopy Development and Management of Mango

Mango (*Mangifera indica*) is one of the least professionally grown fruit crop globally. It is popularly known as 'King of Fruits' belongs to family Anacardiaceae. It is being cultivated for at least 4000 years (Crane, 2008). As per estimate total world mango production is 40 million tones. India ranks first in area and production, accounts 40% of total world's mango production, other important countries are China, Kenya, Thailand, Brazil, Indonesia, Pakistan, Mexico and Bangladesh. India produces 19.68 million tones of mangos from 2.26 million hectare area with productivity of 8.7 t/ha (Anonymous, 2017). United States and Europe are the main markets for imported mangoes. India exports mangos to mainly United Arab Emirates and other countries in the middle east (Balyan *et al.,* 2015).

Export of mango and other fresh fruits have been amplified significantly and increased by 2.6 times in last decade. Leading mango exporting countries are India, Thailand, Brazil, Peru, Netherland, and Pakistan. Leading mango importing countries are UAE, USA, Saudi Arabia, UK, France, Germany, and Malaysia. Only 3.4% (1.65 million tons) of world mango production is traded. India's share in the world trade of mango is around 15 per cent. It is commercially cultivated in Andhra Pradesh, West Bengal, Karnataka, Kerala, Bihar, Uttar Pradesh, Uttarakhand, Maharashtra and Gujarat. In Uttar Pradesh it is cultivated over an area of 2.64 lakh hectare with 4.5 million tones production (Anonymous, 2017). Usually mango gives low yield (8.7 tones/ha) as the tree is huge in size and planted at a wider spacing based on its ultimate size. The main reasons of the low productivity are alternate bearing, mango malformation, fruit drop, insect pest and disease problems and dense and overlapping canopies. A large number of mango orchards in India are becoming old and tree canopies have gone dense and overcrowded. Light interception, is a function of tree spacing, canopy density and height, it is a primary consideration in an orchard designing. In old overcrowded mango orchards, light interception and utilization by photosynthetic surface of orchard is reduced. Jackson (1980) described a close relationship between light interception, photosynthesis and yield in fruit tree orchard. Unfavorable effect of low light interception and utilization on productive physiology of fruit trees is well known.

Rootstock for canopy management

For successful high density planting system in mango, vigour controlling rootstock is pre-requisite. Singh (1976) reported that 'Dashehari' on 'Vellaikulumban' was smallest both in height and spread but on Dashehari it grows vigours. Samaddar and Chakravarti (1988) reported that 'Olour' rootstock imparted dwarfing to 'Himsagar' and 'Langra' cultivar. Singh and Singh (2004) reported that 'Latra' rootstock imparted dwarfing effect to 'Bombai' cultivar. 'Eldon' was proved a dwarfing rootstock for 'Parvin' and 'Tommy Atkins' varieties.

Photosynthesis and light interception

Productivity of tree is depend on the capture of sunlight by the canopy and translocation of photosynthetic to the developing crop. Fruit development is directly related to number of leaves supporting fruits. Photosynthesis depends on the distribution of light and availability of nitrogen within plant. Plants tailored nitrogen resources within the canopy to enhance photosynthesis in location that are exposed to good illumination.

Photosynthetic capacity, carbohydrate amount and nitrogen concentration of leaves were measured in the different parts of mango trees (Urban *et al*., 2003). They reported that the concentration of nitrogen and total non-structural carbohydrates on a leaf area basis increased linearly with incidence of light level. Durand (1997) defined the relationship between light interception and tree architecture, the amount of light penetrating with the canopy decreased the proportion to the leaf area index (LAI)

Light is an important factor for flower bud induction, fruit size and color development. It has a role in fruit development through carbohydrate synthesis. While, increased assimilates in the shoot is a pre-requisite for the flowering of mango (Chacko and Ananthanarayan, 1982), failure of the flowering in mango trees with dense canopies (Burondkar and Gunjate, 1991) and opening of canopies through pruning (Madhavarao and Shanmugavelu, 1976) and in shoot growth retardation (Rameswar, 1989) support indirectly the role of light in fruit bud formation of the mango. Canopy architecture is a natural expression of the genetic make-up of the tree; genotypes vary in their canopy size and shape. Certain scion growth controlling rootstock and interstock like 'Vellaikolumban' (dwarfing rootstock) for HDP of Alphanso (Kurain *et al*., 1996) and 'Esmeralda' on 'Ataulfo' scion as inter-stock has been reported from Mexico (Valdivia, 1999) could be utilized to restrict the mango canopy size. However, the tree size and shape of canopy may also be manipulated through other means which are given here under.

Tree architectural engineering

Commercial mango orchards are planted at 10-12 meter distance (100-69 tree/ha) or an area of 100 to 256 m^2 area is required for growth of the individual tree. After planting, it is imperative to develop effective canopy. Initial branching at the height of 60-75 cm is permitted by heading back. Train the young plants and allow 3-4 horizontal branches to grow, upright growing shoot must be removed and allow only horizontal growing scaffolds. Tip pruning is done to encourage canopy development and ultimate fruit bearing. Allow plants to grow and develop canopy for 3-4 years also. The horizontal scaffolds are stronger and it flowers and fruits early. Short twigs within the canopy produce a complex, compact and strong structure for fruiting shoots.

High density planting system

High density planting in mango was first developed to register high yield without using dwarfing rootstock (Ram and Sirohi 1988, 1989; Ram, 1996). This resulted not only high yield and early returns but also control alternate bearing and ease in mechanical pruning, easy management of pest and disease etc (Ram *et al.*, 1997; Ram and Yadav, 1999). High early production in high density orchard system is the rapid development of leaf canopy and high tree area, which leads to early light interception in the canopy.

According to Jackson (1980), increased in tree density and changes to canopy dimensions may influence the tree growth potential through increased light interception. As dry matter production and yield are related to light interception (Hall *et al.* 1993; Wunsche *et al.*, 1996). High density planting enables earlier cropping, high regular yield, improved farm management practices, leading to higher productivity and profitability.

Mango variety 'Amrapali' could be planted at 2.5 x 2.5m with a density of 1600 plants per hectare (Majumdar and Sharma, 1985). Ram and Sirohi (1985) suggested planting distance 3.0 x 2.5 m for Dashehari which gave 2.43 to 9.43 t yield/ha, compared to 0.19 to 0.32 t /ha under normal planting density (12x12m), after 15-18 years. Ram *et al.*, (2001) found that annual fruit yield increased from 0.30 tonnes/ha in the fourth year to 18.0 t/ha in the 14th year under high density planting (400 trees/ha) and cumulative yield was recorded 74.12 t/ha in 12 years of fruiting. Maximum canopy circumference was recorded in normal density than high density plantation (Ram *et al.*, 2001).

An experimentation on high density planting system was carried out at Meerut U.P. on Dashehari at 10 x 10 and 3 x 3 m spacing revealed that 2.23 to 10.98 t/ha yield was recorded in 5th and 11th years after planting. HDP resulted 10 times more yield (Rajbhar *et al.*, 2016). Training system can change light distribution in the canopy and how it is intercepted across the orchard floor in other fruit orchards (Lauri and Corelli Grapadelli, 2014).

Ibell *et al.* (2018) studied the effect of planting system and training on light interception in mango in North Queensland, as the canopy volume increased; light interception reached a maximum between 61 and 68%. Planting density at 8 x 6 m (208 plants/ha), medium density (6 x 4m), (417 plants/ha) initially density and cultivar had significant effect on light interception; they also observed significant positive relationship between light interception and canopy volume (Ferreira de Sousa, *et al.*, 2012).

'Tommy Atkins' planted at 8 x 5m, 7 x 4 m, 6 x 3m, 5 x 2m and 4 x 2m, plant growth and yield recorded after 7-8 years, above 555 plants/hectare, a significant reduction was observed in mango tree growth and there was also decreased in flowering, fruit yield per plant and per hectare. However, planting density up to 357 plant/hectare inspite of decreasing plant growth and fruit yield per tree but increased in yield/hectare was 30% more as compared to control.

Fig. 15: Mango Variety 'Amrapali' in Full Bearing (Age 5 Years), 3x3 Distance

Fig. 16: Canopy Development in Young Orchard

Fig. 17: Amrapali form Irrgular Canopy

Fig. 18: HDP Orchard of 'Amrapali' Bearing after 4 Years

Table: Yield in mango varieties planted at 3 x 1 m (Oosthuyse, 2018)

Variety	Age	Tree density	Fruit no per Tree	Av. Fruit wt (kg)	Yield kg per tree
Tommy Atkins	1	3,333	10	0.4 kg	13,33
	2		15	0.35	17.44
	3		20	0.35	23.33
	4		25	0.30	24.99
	5		30	0.30	30.0
	6		35	0.25	39.16
Kent	1	3,333	10	0.4	13.33
	2		15	0.35	17.50
	3		20	0.35	23.33
	4		25	0.30	25.0
	5		30	0.30	30.0
	6		35	0.25	29.16

Developing canopy in newly planted orchard

After well establishment of the plants in field which has attained 1m height required to initiate branching. First heading the apical bud on each new terminal shoots once the shoot had fully matured. The tipping process was carried out after every flush. This will hastened the tree development and ultimately compact canopy with vigours terminal shoots.

Tree architecture development

Training is an important tools to develop tree canopy during the initial 2-3 years. This is done with a view to providing a good framework for the future so that the branches are spaced properly, which are able to bear the crop load at the bearing phase. The branches are not encouraged too low on the trunk or too high from the ground level. Fivaz and Stassen (1996) studied the role of training system in maintaining the high density orchards. They observed that the control trees had the highest yield/tree, but central leader trees had the highest yields per m^3 of the tree volume. In order to control tree size and to get maximum production needs appropriate training of the tree at early stage. The trees are allowed to grow up to 4-5 m height by removal of multiple central leaders within the canopy of young tree. In young tree balanced canopy development can be obtained by removal of vertical growing shoots and thus branch angles of the horizontal shoots are improved. The resultant shoots are then headed back when they attained 50-60 cm length. This treatment resulted, a complete canopy structure with an increased number of short shoots (Campbell and Wasielewski, 1999). Increased in branchs number resulted more fruiting shoots in young trees which encourages precocity in flowering and fruiting and also are pre-requisite for high density planting system. Tatura trellis system of training has been found appropriate for low vigour mango cultivars in Australia.

Fig. 19. Training of Young Mango Plant

Pruning

Tree under high density system almost attained the optimum dimension and canopy expansion but slowed down when bearing starts. Canopy expansion is pronounced in off year of the fruiting, due to poor flowering and fruiting but frequent flushing. Pruning strategies for mango should based on requirements of the tree. Annual pruning for flowering management and tree size control; reshaping intermediate sized trees to smaller and more manageable size and complete rejuvenation of large trees which are not productive due to their size and height.

Mango trees experince periodic flushes of vegetative or reproductive shoots initiating on apical or lateral buds from the terminal resting stem. Term stem refers to the vegetative shoots that have become quiescent to become the first intercalary unit at the terminal (Davenport and Nunez-Eli sea, 1997). A shoot usually produces 12 nodes that bears leaves if it is vegetative state and produces lateral inflorescence if reproductive.

Lal *et al.* (2000) reported that increase in the yield of pruned mango trees than the un-pruned due to increase in photosynthetic surface and increased availability of sun light in newly developed canopy. Durand (1997) also found that facilitating light penetration in the canopy during fruit growth lead to increased in yield. Shoot decapitation in the last week of June to 1st week of July followed by 1% urea, after harvesting increasesed new shoot production depending on temperature (Oosthuyse, 1993; Ram, 1999). On year 'Dashehari' mango tree pruned in July, August and December, increased the panicles on shoots, hermaphrodite flowers, and controls in malformation problems, fruit yield and quality. However, off year trees need to be pruned to increase new shoot production and more vegetative growth (Swaroop *et al.*, 2001). Mango pruning in July-August, increased emergence of more number of shoots in the month after pruning which flower in the same 'on' year and produce enough vegetative growth to flower and fruit in the following off' season, which may be helpful to solve the problem of alternate bearing (Swaroop *et al.*, 2001). Shanmugavelu and Selvarajan (1985) observed an increased fruiting in Himayunddin, 'Rumani' and 'Kale' varieties of mango. Pruning in mango was favorable for the flowering by redistribution of endogenous hormones (Madhavarao and Shanmugavelu, 1975) and increasing the total phenolic contents in shoots (Chacko, 1968). The incidence of mango malformation was also reduced by reducing the foliage density by pruning (Sharma, 1953). The pruning increased light penetration in the inner portion of the canopy, resulted, more shoot emergence, panicle formation and reduced malformation problem. Pruned mango cultivar 'Tommy Atkins' had 1/4th of the centre of the canopy removed during March; penetration of photosynthetic photon flux was generally greater in canopies of pruned than in non-pruned trees. The pruned and non-pruned trees have no significant effect on specific leaf weight, however, total leaf chlorophyll content was greatest for pruned trees during November, but similar for pruned and non-pruned trees during April and July in cultivars 'Tommy Atkins' (Schaffer and Gaye, 1989). Pruning of 'Tommy Atkins' after harvest caused marked reduction in yield as compared to control, however, fruit from pruned tree had a better external color than fruits from control (Fivaz and Stassen, 1996). Hence, pruning is an important tools to improve the productivity of old and unproductive mango orchard due to overcrowding canopy by way of rebuilding canopies.

Tip pruning

Pruning of terminal stem not less than 1 cm diameter of any length from tip of the stem, which is dark green wood on 2-3 inter calary unit is called tip pruning. Pinching is also a type of tip pruning where apical portion of plant is terminated with aim to attain more growth. Pathania *et al.* (2000) reported suppressive as well as beneficial effects of the pinching in terms of flowering and increasing in number of flowering surface (flowering wood). Soudager *et al* (2018), reported that tip pruning up to 2 leaves induced early flowering and resulted in earliest harvest in Alphonso mango. Pinching and tip pruning forced rapid initiation of dormant lateral stem buds to form lateral shoots. Tip pruning in mango is carried out for several purposes viz to stimulate branching flushes of lateral shoots through repeated pruning in young trees to from a dense spreading canopy in which flowering starts 1-2 years earlier than normal. It is also done to remove growth inhibiting panicle.

Tip pruning shorten the Juvenility of young trees as most the mango saplings in nursery have only main stem with only few lateral or no laterals. After planting if frequent tip pruning carried out, forces trees to have development of lateral shoots forming 4-7 shoots per stem. If pruning is repeated every 3 months, full canopy develops earliest. The flush frequency reduces and shoots get mature for flowering and fruiting. In the northern hemisphere, the final pruning must be carried out in August or early September for the resulting stem to achieve sufficient maturity by the time of natural flowering in Jan-Feb.

Pruning for shape or formation pruning

The objectives of shape pruning are to reduce the dimension of the canopy which has become overcrowded. This kind of pruning involves cutting of branches ranging from 2-10 cm in diameter, according to size of the tree. A severely pruned tree tends to attain original shape and size. Frequent tip pruning of the resulting shoot reduces the flush frequency back to normal usually by one year resulting in resumption of flowering. This kind of tree can be maintained in desired shape and height, for many years using annual tip pruning.

Pruning for rejuvenation

The foremost reason of low productivity of mango in India is more frequency of old, unproductive and senile orchard with unreachable canopy. Such orchard has poor fruiting on the upper periphery of the canopy. As middle and inner most parts of canopy did not receive adequate amount of solar radiation Experiment conducted in various Institutes revealed that such orchard can be rejuvenated employing severe pruning technique. Rejuvenation of mango orchard through pruning has already been standardized in traditionally grown mango cultivars/orchard (Burondkar *et al*., 2000; Lal *et al*., 2000, Mallick *et al*., 1985; Shinde *et al*., 2002). Amrapali planted at 5x5m in HDP, rejuvenated ofter 24 years, employed pruning height 1m, 1.5m and 2.0 m height from ground with primary branch pruning at 60cm and 120 cm. Best results obtained with 1.0m pruning height with 60 cm primary shoot length and no control on length of secondary shoots found most appropriate treatments to attain an adequate canopy architecture (Das and Rana, 2013). Burondker *et al.* (2000) carried out rejuvenation pruning of 34 year old Alphonso trees in 1993-95 and found that rejuvenated trees treated with paclobutrazol 2.5, 5.0, 7.5 and 10g/ tree, flowering advanced in all the concentration with 69.17 kg – 96.65 kg fruit yield per tree. The soil residue of 7.5 and 10g per tree was also noted in subsequent year.

Fig. 20. Twenty Five Years Mango Variety 'Dashehari in Bering after Rejuvenation Spacing 2.5x2.5 m.

Fig. 21.

Though mango is evergreen tree hardly require regular pruning, the diseased, dried and criss-cross branches are removed. However, reduction in flower bud induction, fruit size and color development, are current emerging issues because of poor light levels due to overcrowding within and between the tree canopies (Jackson, 1980; Flore, 1994; Whiley and Schaffer, 1994). Favorable effects of different intensities of pruning in mango on light interception, chlorophyll content in leaves of pruned trees (Schaffer and Gaye, 1989) and yield (Rao and Shanmugavelu, 1976). Mango pruning is necessary due to frequently flushing. In the tropics and subtropics climatic conditions, tree size and productivity of commercial orchards must be maintained. Pruning employed in mango trees, increased the penetration of light through the canopy and increased the rate of photosynthesis (Schaffer and Gaye, 1989 b) Pratap *et al*., 2003; Sharma *et al.*, 2006).

Sharma and Singh (2006) studied the pruning effect on the productivity of 16 years old 'Amrapali' variety of mango in New Delhi, the branches were tipped to remove newly emerging shoot or the branches were cut at 10, 20 and 30 cm of new growth with control tree. The result showed that top section of the canopy have maximum yield than lower most and middle canopy. Tipping and pruning increased the number of inflorescence and yield in all the section of the canopy. Then high density orchard of 19-20 years old of 'Dashehari' at ICAR-CISH, Lucknow showed that after severe pruning (severely pruned trees at 2.0m to 2.5m pruning height started bearing fruits earliest than 1.5 meter pruning height (Anonymous, 2020). However, flowering and fruiting started in 3rd-4th

year in 1.5m pruning height. Sharma *et al.* (2005) reported that in 'Amrapali' pruning at 10-15 cm of tip of branches improved light penetration. Davenport, (2006) reported that tip pruning can be used to encourage frequent flushing and branching of young trees to bring them into commercial production. Formation pruning gives shapes to the trees in a flowering management programme. He further reported that severe pruning coupled with subsequent tip pruning of huge, non productive trees, restored the productivity. Majumdar *et al.* (1982) reported more yield in high density orchard during initial year in 1600 plants/hectare against low density, it is due to greater land use efficiency during early years. It is important to impose containment pruning to control canopy for regular production. Charnivi Chit and Tongumpai (1991) were able to reduce canopy spread in mango orchard in 2.5x2.5m spacing which was subjected to severe pruning along with paclobutrazol application resulted flowering and fruiting profusely.

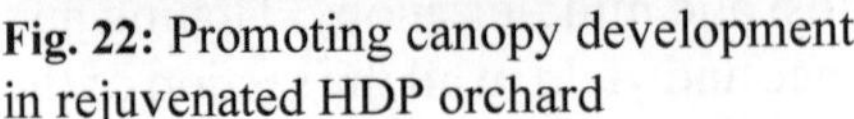

Fig. 22: Promoting canopy development in rejuvenated HDP orchard

Fig. 23: Containing canopy spread through annual pruning

Most of the old mango orchards have been planted on traditional spacing with no proper canopy management system. Such tree has become tall which is beyond reach resulting low light interception in-side the canopy and hence, fruiting takes place on the canopy surface. Commonly mango forms varied canopy shapes i.e. dome shape, umbrella shape, semi-spherical, open, spherical and scattered canopy. Amrapali variety do not form compact canopy in other ward it form scattered canopy (irregular canopy). Semi-spherical canopy is usually common in mango with many small openings. Rejuvenation pruning has been found successful to rebuild the canopy to regain productivity of old, senile and unproductive mango orchard.

5

Canopy Development and Management of Grape

Grape is one of the important fruit, crops being grown in area of 79.6 thousand hectare with annual production 1878.3 thousand MT. India is a major exporting country of grapes, it exported fresh grapes 24633.79 MT (2018-19) (Anonymous, 2018). India exports grape mainly to Netherland, Russia, United Kingdom, Bangladesh and Germany. About 80% grape production comes from Maharashtra followed by Karnataka and Tamil Nadu. Fruit contains 20% sugar in easily digestible form, also rich in calcium and phosphorus. More than, 82% grape used for wine making 10% for raisin making and rest for table purpose world over. However, grape is mostly consumed as fresh in India and only a limited amount is used for processing purpose.

Over past few decades advancements have been made in vineyard design. Tree architectural techniques have improved grape quality production. 'Vertical Shoot Positioning' (VSP) and 'Y' Trellis are the most common tree architecture used in Southern Brazil. In VSP system vine shoots are trained upward vertical, narrow curtain with the fruiting zone below. VSP (Vertical Shoot Positioning) tree architecture consists of 4-6 levels of wires. VSP training system is good for small growing canopy varieties. VSP training systems helps to thinning, leaf and leaf positioning. This system has efficient coverage of spray materials and ideal for mechanization in pruning and other cultural operations.

The physico-chemical constituents of the grape berry depend upon the stage of development, variety and rootstock. Different environmental conditions viz. soil structure and composition, moisture, temperature and light have their own influence. Vine do not grow well under humid conditions, as such it provide continual condition for disease development. A bright sunny days helps in the development of sugar in berry. High UV radiation affect the pigment synthesis, inhibited leaf and shoot growth, hastened tissue differentiation, lignifications, shoot maturation, ripening and sugar content of berries (Zhakole *et al.*, 1979)

Canopy development is utmost important in weak growing stem to facilitate maximum light interception and penetration. While increased assimilates in the shoot is a pre-requisite for flowering in grape (Thomas and Bernard, 1937).

Light effect on growth yield and quality

Light was found to perform a triggering action in the process of fruit bud differentiation in grape (Baldwin, 1964). The light requirements varies from variety to variety while as 'Riesling' variety of grape require less light intensity and 'Thompson Seedless' requires more light for the fruit bud formation. Higher, light intensities of more than 3600 ft candles and temperature above 35^{0}C are favorable for the bud fruitfulness in 'Thompson Seedless' grape (Buttrose, 1969).

The light utilized by the plants for the photosynthesis corresponds to 400 to 700 nm of the electro-magnetic radiation from the sun. Kriedmann and Smart (1973) have reported that the photosynthesis in grape rapidly increase up to the light intensity of 5,000 ft candles. The light compensation point at which the rate of photosynthesis is just the equal to the rate of respiration in 'Thomson Seedless' grapes is 125 ft candles. It is a well-established fact that the leaves at the light regimes of lower than the compensation points are the liabilities to the plant. A leaf absorb more than 90% of the solar radiation depending upon its thickness (Shikhamany, 2001). In full sunlight in a given locality is 12,000ft candles, the third layers of the leaves in a tree canopy would receive the light at lesser intensity than the compensation point (Smart, 1973).

Though canopy architecture is a natural expression of the genetic constituents of the tree, however, the size and shape of the canopy may be manipulated through various horticultural techniques. Canopy management by accommodation of foliage on the trellising system, shoot positioning, suckering-topping and **shoot positioning-suckering**- topping and leaf removal resulted in the highest yield (Hunter, 2000). Low light during inflorescence development, reduced bud fertility although the bud break was earlier and the rate of development of new shoots was higher in the following season than grape grown in inadequate light. Poor canopy management reduced the bud fertility, fruit quality and increased in flower abscission as well as inflorescence and bunch necrosis (Koblet *et al*., 1997). Canopy management resulted lower susceptibility to stem necrosis and Botrytis infection.

Volschenk and Hunter (2001) studied the effect of seasonal canopy management trained on lengthened perold trellising system. No canopy management (shoot growing in all directions) resulted in over exposure of the bunch zone directly above the cordon, whereas sunlight reflection from the soil was directly reduced. In contrast, canopy management practices resulted more balanced penetration of sunlight into the bunch zone. Air flow through the canopy, was found highest when partial defoliation in combination with suckering and shoot positioning carried out. These practices have highest impact on canopy micro-climate development and canopy appearance. The highest yield was obtained by applying shoot positioning and defoliation on topping.

Grape canopy

Shikhamany (2001) described an ideal canopy model for grape cultivar 'Thompson Seedless'.

Cane number	5 per m^2 (Shikhamany, 1983)
Cane thickness	8 to 10mm (Shikhamany, 1983)
Stem height.	135 cm (Shikhamany, 1983)
Diameter	7.5cm (Shikhamany, 1983)
Cordon length	90 cm (Shikhamany, 1983)
Shoot orientation during the growth season	35^0 to 40^0 with ground surface up to the length of 90cm and parallel to the ground, surface beyond 90cm (Shikhamany, 1983)
Leaf number per Bearing shoots	12 to 15 (Chittirai Chelvan *et al* 1985)

Tree architecture

Training is a potential tools to develop good canopy architecture of plant with weak stem. Bower system of the training has been found to be the best in tropics, throughout the world. Although it is an expensive training system associated with the reduced light and temperature in vine canopy, increased humidity and disease incidence, it is inevitable for fully exploiting the productive potential of the grape vine in tropical region, where phenomenon of the apical dominance is more pronounced.

Wunderer and Mayer (1994) compared high trellis, single wire trellis, single Geneva double curtain (GDC) and double GDC training system. Both GDC training systems resulted in the highest average fruit weight, followed by high trellis and single wire trellis. Open espalier-type tree architecture is more suitable for mechanized canopy management than single wire or GDC systems.

It is possible to develop as many as 10 shoots per m^2 by subdividing the apical growing in horizontal position. Vertical canopies have been found best for utilization of the light and the minimization of building up of high humidity in side canopy, do not have provision to increase the number of fruiting units per unit area. The appropriate method to manage the canopy in grape vines is to develop diageotropic canopies and increased the fruitfulness of the buds and consequently the cluster, cane ratio (Shikhamany, 1988).

Bower or arbour or pergola architecture

Vigours growing grape vines are best trained on this architecture. In this system the vines are spread over the criss-crossed network of wires. It is the most expensive architecture system among all and provide good protection of the crop against hot desiccating wind. Apart from being costly, pruning, training and spraying operation become difficult. The vigours growing vine varieties are trained in this architecture.

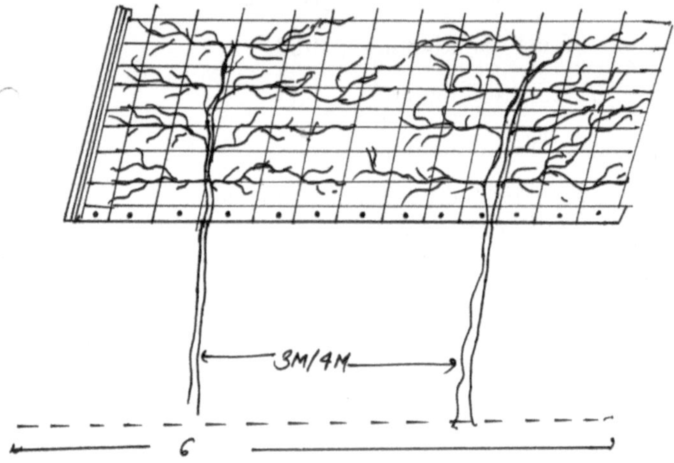

Fig. 24. Bower or pergola architecture

Kniffen systems and overhead trellis of telephone systems

This type of architecture is popular in north India due to less vegetative growth and is also suitable for less vigor growing vines. It is an improvement over Bower system in respect to ventilation and light penetration. Vines are trained horizontally along the GI trellis wires fitted on the Iron supports. The main scaffolds are allowed to grow 1.5-2 m high. Each vine has 2-3 wires on which 4-6 arms develops at interval of 0.3 meter. Cane pruning varieties i.e. Pusa Seedless and 'Thompson Seedless' are suitable on this system

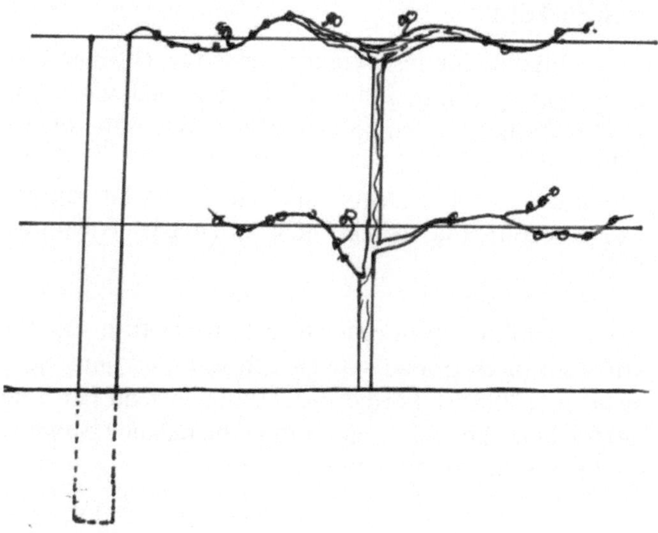

Fig. 25. Two-wire Vertical, vines (Kniffeen system)

Telephone type architecture

It is also known as 'T' trellis, semi- vigours grape vines are trained on telephone type architecture. It is an improvement over 'bower' system and is less expensive than Bower. 'T' trellis facilitates better ventilation and light penetration in the canopy. However, yields in this system are low in comparison to the 'bower' system due to less number of canes per unit area. The vines are allowed to grow straight up to a height of 1.5-1.75 m on which 2-3 primary scaffolds are developed. On each of the primary scaffald branch short secondaries of 30-45 cm are developed on both sides of the primary scaffalds and fruiting canes are developed on these short secondaries.

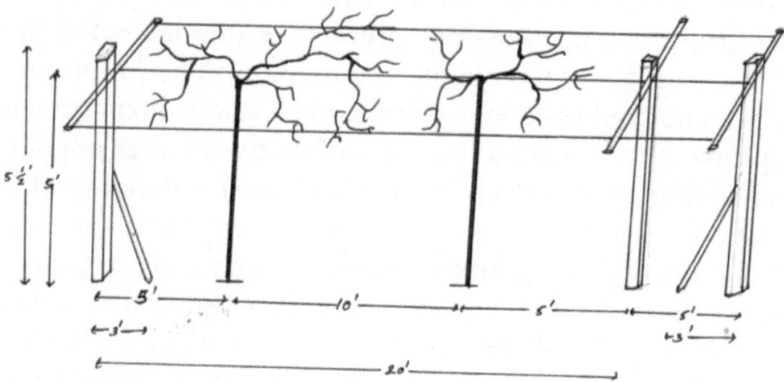

Fig.26. Telephone type architecture

Head or single stake architecture

This type of architecture is suitable for less vigours growing cultivar like Beauty Seedless. In this system of training the single stemmed vines are allowed to grow initially with the help of support or stakes. When the plants grow up to 1.2 m, the trunk is cut back to produce side shoots, 4 laterals in all the directions are allowed to grow, on which secondary and tertiary laterals are allowed. After 3-4 years, vine become like a dwarf bush and needs no staking.

Crop load

Optimizing the crop load is the important in order to get proper fruit size for reducing the input cost. In want of crop load standard fruit quality and yield both affected. Somkuwar *et al*., (2020) reported that 60 bunch load per vine with 5.43 kg increased yield over 40 bunches maintaining the balance between yield and quality was found better in 'Manjari Naveen' grape.

Pruning

Pruning refers to the judicious removal of any plant parts to establish and maintain desired vine shape with aim to increase productivity and easing in various cultural operations. Distribution of proper amount of bearing wood over the vines and to regulate the crop for maintaining the vitality of vine and consistent productivity.

Pruning and shoot topping are the regular practices to shape the canopy and to promote the fruiting and ripening in grape. Pruning is done once in north India during the month of January, while in south India, twice a year, once in summer and again in winter. Summer pruning is done during March-April in the states of Andhra Pradesh, Karnataka and Maharashtra but in June in Tamil Nadu in such case back pruning or growth pruning is done.

Sommer (1995) recorded improvement in fruit vine quality but reduced in yield by mechanized crop thinning. Summer pruning results indicated that yields were reduced but fruit and wine quality improved. Minimum pruning produced substantially higher yield but delayed ripening. Canopy management also includes techniques to open up dense leafy canopies by summer pruning, leaf removal and trellis change shoot positioning and cluster thinning also. Removing within canopy shade by canopy management technique can increased both yield and quality in vigours vineyards. Application of canopy management techniques can decrease the yield gap between distinguished and ordinary vineyards (Smart, 1993). Thinning out and ethephon application advanced harvest dates by 5-8 and 2-3 days respectively, berry weight was increased by thinning but not affected by leaf removal or ethephon application (Fitzgerald and Patterson, 1994)

The plant spacing has direct relation with light interception in the canopy. The number of leaf layers, light intensity and air flow generally decreased with closer spacing, whereas, relative humidity increased. The lower photosynthetic activities of closely spaced vines were accompanied by increased transpirational water loss and are mainly attributed to less favorable canopy micro-climate. Yield per hectare grape was recorded higher in closely spaced (1x1m, 1x0.5m) vines, required significantly higher inputs for canopy management, harvesting and pruning. Medium spaced vines (2 x 2m, 2 x 1m), considering the land utilization, vine performance, vine quality as well as labor use efficiency were recorded optimally (Hunter, 1998). Nutrition has great influence on canopy growth, vigor and canopy volume. Nitrogen favors the increment in tree vigor and volume, while potassium reduces them. Shikhamany and Chadha (1993) reported that control in canopy vigor, shoot orientation and foliage density with the application of potassic fertilizers during the growth season under double pruning and single cropping system in tropical climatic zone in grape cultivars 'Thompson Seedless' grape.

Canopy management for disease control

Disease intensity in grape depends on the climatic condition and canopy architecture. Bregglio *et al.* (2011) suggested that in addition to genetic resistance, the use of a plant architecture that produces less favorable micro-climate condition for fungal infection could significantly reduce disease.

It has been observed that grape vine with dense canopies, thin skin, tight clusters are much susceptible to Botrytis bunch rot (Pezet *et al.* 2003; Ky *et al.*, 2012, Muvde *et al.*, 2012). Downy mildew and botrytis bunch rot occurrence are mostly related to environmental conditions like humidity, temperature and light.

Canopy management practices also influenced the disease occurrence. Bunch rot decreased greatly by leaf removal treatment. In Monterey County (California), leaf removal treatment in 1984 and 1985 reduced disease incidence from 11.9 to 1.8 and from 55.0 to 23.9 respectively (Gubler *et al.*, 1987)

Shoot thinning (removal of 12-15 shoots two weeks after fruit (set) along with shoot tipping (shortly before veraison) decreased the development of stalk necrosis, but shoot tipping alone did not. The effect on stalk necrosis development were related to light interception, which was the highest in combination of shoot thinning and tipping. Further, light interception can also be increased by shoot thinning and tying of shoots (Perez Harvey *et al.*, 1987)

Pereira de Bem *et al.* (2015) demonstrated that canopy manipulation in the vine yard can have significant effect on the development of downy mildew and *Botrytis* bunch rot on leaves and fruit clusters of grapevine they further reported that incidence and severity of downy mildew and *Botrytis* bunch rot were greater for Y Trellis training system than VSP (Vertical Shoot Positioning) architecture and VSP training system was most efficient to prevent or suppress disease development.

6

Canopy Development and Management of Banana

Banana and plantain (*Musa* spp.) are one of the most important commercial food crops especially in tropical regions. Banana by virtue of its multiple uses is popularly known as 'Kalpataru'. Apart from fresh consumption as desert fruit, some types are also used for culinary purposes. It has therapeutic value; ripe banana is easily digestible and has vitamins and minerals. Banana is known as wholesome fruit as it provides more balanced diet than any other fruits. It is rich source of carbohydrates, the source of energy.

India is the largest producer of Banana in world. Producing 29,20000 MT with average productivity 21.00 t/ha from an area of 1,39 000 hectare area (2017-18) (Anonymous, 2018). India, exported 188,22000 MT banana the value of Rs 189,994.86 lakhs. Banana is the largest fruit crop in India, forming nearly 32 per cent of total fruit production (Singh and Chadha, 1996). Banana production has increased by 209 per cent during the past 20 years due to advancement in cultivar development, disease and pest control. However, there is further scope to increase yield of uniform color fruit by proper canopy management and provision of light penetration.

Banana and plantains are grown mostly by small holder and plays an important socio-economic role in many developing countries of the tropics. It is particularly suited to intercropping system and to mixed farming system. Banana is often used as a ground shade and nurse crop for a range of shade-loving plants along with cocoa, coffee, black pepper, coconut and nutmeg. In most banana-growing regions, solar radiation is abundant and their productivity largely depends upon the efficient utilization of the resource. In multistory cropping system, banana is grown to harness maximum light, land and nutrient availability.

High density planting

High density planting of banana in new concept and new field of orcharding system, it is the way to increase productivity without affecting fruit quality. Lot of research work have been carried out in different parts of the world in varied agro climatic conditions.

Gogoi (2015) planted 2-3 suckers per pit and recorded more number of functional leaves as compare to control. All the plants under HDP recorded reduced bunch weight, number of hands per bunch, fingers per hands, length and yield (80.23 t/ha) was found in 3 sucker/hill at 2 x 3 m spacing with 50% recommended fertilizer dose was best in Assam agro climatic conditions. High density planting in banana has the advantage of utilization of better solar radiation, checking excessive vegetative growth, reducing competition between fruits and alternate sinks such as supping structures, reduction in the distance the assimilates have to be translocated from source to sink and all these lead to higher productivity. However, to determine the optimum density for a given level of plantation, vigor canopy characteristics such as leaf area index and transmission of photo synthetically active radiation (PAR) can be used to correlate with yield (Stover, 1984).

Canopy management

It is pre-requisite to address canopy management aspect for success of high density plantation in banana. Training and pruning practices are important tool to manage the canopy. Productivity in banana is governed by source, sink relationship. Leaf area Index may be used as management tools (Turner, 1998). High density planting of banana has direct impact on plant growth, yield attributes, leaf area index and leaf numbers (Nalina *et al*., 2000; Thippesha *et al.*, 2005).

In banana, canopy composition depends on the plant density and spacing (Rodriquez *et al*., 2007). Leaf area index was directly proportional to the planting density. Maximum photo synthetically active radiation is intercepted by increased planting density. Debnath *et al*. (2015) recorded optimum leaf area index (5.50) for high density planting of 'Martman' planted with popular density 5000 plants per hectare at 2 x 3 m spacing with 3 suckers per pit.

Populations (mother plant plus one daughter) in permanent banana plantation should be based on leaf are index (LAI=m^2 foliage per m^2 plantation) and photo synthetically active radiation (PAR) transmitted through canopy. At the same planting density, ‘Grand Nain’ had an LAI 20% below that of 'Valery' in the light clay soil ‘Valery’ had LAI 15% and Grand Nain 20% below that on a loam soil. In 'Valery' cultivar leaf area Index, where growth and production were very good (43 kg average bunch weight with 15.5 leaves per plant) the leaf area index was 5.1 and 6.1 for populations of 1700 and 1900 plants per ha, respectively. In less productive plot (37 kg bunch weight, 12.5 leaves per plant), leaf area index at populations of 1700 and 2100/ha was 3.5 and 4.5 respectively. In the first leaf area only 2.3% of PAR was transmitted to the ground and in second leaf area 20 and 13% at two populations, respectively (Stover, 1984).

Light interception, soil fertility, climatic conditions, soil moisture content etc, are important points to consider for laying out of high density plantations. Cultivars planted in different densities resulted fruit yield 86.6-174.39 kg per plant at 4444-10,000 plants per hectare. In plant population of 4,444 plant per ha total fruits yield recorded 120 t/ha (Awasthi and Mehta, 2000). The fruit quality and yield of banana vary with plant height, leaf emergence, leaf area, suckering habit, bunch weight and number of hands per bunch. The principle advantage of high density plantation is of minimum weed growth, better utilization of solar radiation, less pseudo-stem growth, lesser leaf number, however larger leaf size and low suckering delays in flowering and fruit maturity. Some cultivars like 'Robusta', 'Dwarf Cavendish' and 'Lacatan' are good for HDP and high yield.

Pruning

Pruning of surplus leaf is a common operation in banana cultivation. Leaf pruning improve light penetration and reduce disease spreading through old and senescent leaves. The micro- climate also changed by leaf removal especially light and temperature. For optimum crop production minimum of 12 leaves are required to be retained.

High yields of quality fruits in banana is possible by increasing planting density, canopy management practices, use of genetically dwarf cultivars etc. Some growth controlling horticultural techniques can also be adopted to control canopy of banana under different planting system.

7

Canopy Development and Management of Guava

The guava (*Psidium guajava* L.) also known as 'the apple of tropics' belongs to family Myrtaceae (Hayes, 1974) originated from Tropical South America stretching from Mexico to Peru and introduced in India by Portuguese during 17th century (Bose and Mitra, 2001). Guava is an evergreen fruit in nature, well adopted to a wide range of soil and agro-climates and acclaimed as 'super fruit, due to high nutraceutical properties and in reach of poor people it has great demand. It is successfully grown in saline, alkaline and acidic soil. Guava fruits are used for fresh table purpose as well being cultivated over an area of 260,000 ha with 38,26,000 MT fruit produced (Anonymous, 2018).

According to Bruton, in the early years of seventeenth century it was found growing in India. Guava is one of the richest and cheapest sources of vitamin C (100 to 120 mg per 00gff). The ideal canopy shape should be such that, it may meet out as many as possible principles. The perfect ideotype of guava is canopy size should be dwarf, spreading, larger trunk and open. The tree size can be controlled by use of dwarf imparting rootstocks like *Psidium friedrichsthalianum* (Edward and Shankar (1964) and Aneuploid 82 (Sharma *et al*., 1942). While as, triploid guava was found to increase tree vigor of Allahabad Safeda cultivar (Prasad, 1966).

Guava bears flower and fruits on the current season of recently matured shoot, either from lateral buds on older wood or at the shoot terminal. Thakre *et al*., (2013) reported that increased in number of current season's shoot significantly influenced the productivity of the tree (Singh, 1985). Hence, management of new current shoots significantly influences productivity of the plant. Guava responds well to different type of pruning practice. Shoot pruning is also helpful in reducing the tree size and proving the fruit size and quality. Different kind of pruning can be used for managing the canopy architecture under different planting densities.

Flowers are produced in the axil of branches of current season's growth and are white in color. Fruit is a berry, globose, ovoid or pyriform in shape. Guava bear's flower and fruit round the year but winter, summer and rainy are the main fruit bearing seasons. Pruning is an important technique for increasing the fruiting point in guava trees; however, it influenced the particular phenological stage. Pruning in fruit trees starts with their planting in order to give particular framework as per tree morphological nature, which enabled trees to bear fruit load. Pruning in fruit trees are drastic operation, which directly or indirectly influences many physiological activities.

Pruning also helps to maintain canopy spread and height, easy spraying, fruit picking and thinning of flowers and fruits. Pruning also prevent excess fruiting improve fruit quality and facilitates light penetration into the inner parts of the canopy, which ultimately improves the fruit color development.

Pruning results are not same for all the cases, because pruning effect depends on tree vigour, tree species, morphological features, cultivars, growing conditions and other factors. (Gardener *et al.,* 1922). Pruning stimulates the growth of the young, vigorous trees grown in favorable conditions more than that of older on dwarfing rootstocks as flowers and fruits are borne on current season's growth and light annual pruning encouraged new shoots growth after fruit harvest (Mika, 1982).

Growth regulating plant hormone might be playing key role in performing the communication systems by switching certain genes. The hormonal status got changed when shoot apex is removed, resulting lateral buds stimulation for growth, branching is induced, photosynthetic activity in basal leaves enhanced dry matter production. Dry matter partitioning is changed in such manner that the tree is able to re-growth the removed portion quickly.

Pruning time on tree growth

Pruning is important tool for limiting the tree size and improving the fruit quality, it is practiced after fruit harvest in different fruit crops under high density planting system. The light pruned trees have large number of flower buds emergence, while as plants subjected to severe pruning resulted least numbers of flower bud growth, ultimately less fruits yield. Jadhav *et al.* (2002) observed that delay in pruning from 25 April to 25 June, in 'Sardar', resulted increased in the shoot length, number of flowers, fruit numbers per shoot, average weight of fruits and yield, whereas, Chandra and Govind (1995) reported optimum pruning time is February in north-east India. Aswathy and Arumugam (2017) reported mid March best time. Shoots pruning resulted increased in yield and yield attributing like fruit weight and fruit yield per tree in Lucknow-49 guava at South India.

The phenological development of guava tree was greatly influenced by the pruning practices and heat condition. Guava tree required approximately 800 to 850 heat units and from 1950 to 2000 heat units from pruning to flowering and flowering to beginning of fruit harvest respectively (Padilla- Ramirez *et. al.* 2012). Enhancement in the winter crop may be due to the exposure of the stress condition by way of pruning during summer, pruning during summer might have brought down the Carbohydrate (CHO) content of the tree to such a level that tree remaining portions have stimulated for vigorous vegetative growth in order to bring the C: N ratio at optimal level.

Pruning for crop regulation

Guava has tendencies to bear fruit at different times of the year, but harvesting of all amount of fruit is not profitable. The quantum of guava production is very high in rainy season (Rathore and Singh 1974; Singh *et al.* 2000), guava bears most crops in the rainy season, however, rainy season guava is poor in quality, insipid in taste and also fruits are affected by many biotic and abiotic stresses. On the contrary winter season fruits gives quality fruits and high monetary profit. This necessitates crop regulation techniques to harvest quality fruits in winter season. Crop regulation has been standardized to induce new vegetative growth during rainy season so that good crop is harvested (Shigeura and Bullock 1976; Singh *et al.,* 2000). Fruit harvested in the winter season are superior in taste and appearance and having good market demand. The winter crops which ripes in October to January are superior in quality and taste.

Table 1: Time of flowering and fruiting in guava

Western India		Northern India	
Flowering	Fruit harvesting	Flowering	Fruit harvesting
Feb- March	July-Sep	Feb- March	July- Aug
June- July	Nov- Jan	Oct- Nov	March- April
Sep- Oct	Feb- April	June- July	Nov-Dec

The basic principal of crop regulation is to manipulate the natural flower and fruiting of guava for the desired season. Required crop regulation practices are done by creating stresses, with-holding irrigation during flowering and fruiting, root exposure, pruning, thinning of flowers or spray of NAA on blooming tree. Severe pruning resulted earliest sprouting of new shoots with flower drop in the rainy season (Mohammed *et al*., 2006). Pruning of current season's growth is effective for crop regulation (Tiwari and Lal 1984; Singh 1980), Dhaliwal *et al*. (1984), reported that 25-52% shoot pruning in 'Sardar' between 20 April to 10 May was found effective to avoid rainy season flowering and encouraged winter crop. One fourth pruning intensity of 'Sardar' can regulate crop yield without affecting fruit quality under HDP, where as pruning in May resulted least rainy season crop and increased the winter season crop (Das *et. al.,* 2018; Jadhao *et. al.,* 1998 and Singh *et. al.,* 2015; Agnihotri *et al*., 2016; Kumar *et al*. 2017).

Table 2: Flowering and fruiting time in eastern and south India

Eastern India		Southern India	
Flowering	Fruit harvesting	Flowering	Fruit harvesting
Feb- March	June- July	Feb- March	June- July
June- July	Oct- Jan	Oct- Nov	Feb- March
Oct- Nov	Feb- April	June- July	Oct-Nov

Lal *et. al.* (2017) and Singh *et. al.* (2018)

Pruning severity

Severe pruning resulted decrease in the leaf area, photosynthesis and translocation of photosynthates to fruits as well as roots and increases in the root, shoot ratio (Casierra-Posada and Fischer 2012; Lotter, 1990).) Which favors the vegetative growth. Fifty per cent shoot pruning in April and July have positive effect on plant height, canopy spread, stem diameter and plant volume, However, more number of new shoots emerged in April pruned in meadow guava orchard. October pruning did not make any positive effect on fruit quality (Shah and Lal 2017). Pruning proved to be successful in rejuvenation for reviving yield and quality attributes (Basu *et al.,* 2007). Heading back and thinning out resulted removal of terminal meristems which is the seat of auxin content while root continue producing cytokinin, which stimulate cell division and shoot growth.

Heading back of young bearing trees in summer helped to improve light penetration inside the tree canopy. Removal of 20 cm shoot tip in early May, enhanced the leaf numbers and leaf size both in rainy and winter season, pruning in, May also increased the fruit size and fruit weight in both season and also improves the fruit quality (Adhikari and Kandel 2015; Joshi, *et al.* 2016). Chandra and Govind (1995) reported that above 70 per cent pruning in high density planting of guava, resulted highest fruit yield.

Pruning in ultra high density planting system

Das *et al.* (2018) conducted pruning experiment in ultra high density orcharding system and found that 80% pruning of canopy in October resulted maximum yield of rainy season crop in 2009-10 and 2011-12 (16.86 t/ha and 25.58 t/ ha respectively), however, pruning of 60% in May resulted maximum yield in winter season. He further reported that 50% shoot pruning in March, May and October found promising without affecting the quality of the fruits. Wang (1987), open centre trained, 4-5 years old trees branch were bended, pinched which indicated that bending of the branches ware most effective in controlling vegetative shoots and increasing yield/tree (21.43kg), while as yield in control recorded low (7.10 kg per tree).

Pruning for yield improvement

Pruning in evergreen tree was unusual practice, accidental damage of branches etc. lead new growth on which flower and fruits borne. Pruning intensity has profound influence on flowering and fruiting, removal of 15-30 cm of branches adversely affected the flower production and ultimately reduced fruit yield per branch (Sheikh and Hullamani 1993; Barar *et al.,* 2007 and Singh *et al.,* 2007) he further added that pruning intensity had positively influenced the fruit weight and volume. High yield (46.63kg/tree) was recorded by pruning at 2 m height in 'Allahabad Safeda' (Singh and Singh 2007; Kumar and Rattanpal, 2010; Mohammed *et al.* 2006; Dalal *et al.,* 2000; Lotter and De 1990; Singh *et al.,* 2012). Increased in the yield of pruned tree was due to increased in the level of photosynthetic active radiation within the tree canopy.

Pruning significantly decreased the fruit set percentage and fruit number per plant during rainy season and subsequently increased during winter season, since the food reserved in rainy season utilized for winter crop growth. Maximum fruit yield (6.53 and 6.23 kg/plant) noted with thinning out of non-fruiting shoot in winter season planted at 2×1m spacing (Nautiyal *et al.* 2016). Dalal *et al.* (2002) observed when 2.0, 3.0 or 4.5 cm thick shoots were pruned increased the fruit set and size. The overall increase in winter crop after removal of rainy season crop in May is due to higher accumulation of assimilates which was not utilized by rainy season crop (Clair-Maczulajtys *et al.,* 1999; Singh *et al.,* 2001). Light and moderate pruning resulted highest number of flower buds and fruits which were ready to harvest in 160-180 days after pruning (Anez 1998; Sheikh and Hulamani (1994).

Fruit quality improvement by pruning

Open central leader of training system has been found good for quality yield (Pereira, 1990), induction of high fruit quality by different pruning levels have influenced significantly on fruit quality. One third shoot pruning in summer has significantly increased fruit yield and quality of rainy season guava and recorded total cumulative yield (57.30kg/tree). Guava pruning treatment also increased the pectin content in fruit; it was higher in winter crop as compared to rainy season (Joshi *et. al.,* 2016; Satyaprakash *et al.,* 2012)). Pruning height has also notable impact on fruit quality, maximum fruit length, diameter, weight, volume, specific gravity, TSS, total sugars and ascorbic acid content were found maximum in 1.5 m pruning height under HDP (2.0x1.5m) (Kohli *et al.* (2017; Singh *et al.* 2001; Joshi *et al.,* 2016). Moderate and light pruning in guava, resulted greater fruit weight, in contrast to heavy pruning (Serrano *et al.,* 2007). In 'Lucknow-49, cultivar, removal of 30 cm of apical shoot in mid March in South India resulted high TSS, ascorbic acid and total sugar content. Similarly 10 cm shoot tip pruning gave highest fruit yield (Suresh and Shakila, 2017; Singh and Dhaliwal 2004; Dalal *et al.* (2000) in contrast severely pruned trees produce smaller fruit and lower TSS and sugar level (Singh *et al.,* 2001). Pruning resulted better quality fruits during the rainy season (Sheikh and Hulamani, 1994; Dasarthy, 1951; Anez, 1998; Sheikh and Hulamani, 1994 and Bajpai *et al.* 1977).

Fig. 27. Pruning in HDP Guava

Fig. 28. Removal of 50-70% Annual Growth and Central Shoot Training

High density planting system has resulted early economic production, more return per unit area, provide efficient use of natural resources. Guava bears on newly emerged shoots, hence, it is well suited for high density planting. Mostly guava is planted at 6-8 meter spacing, however, with the advancement in knowledge, it is being planted at 1x1m, 2x1 m spacing also. At ICAR- CISH, Lucknow, different densities 3 x1.5 m, 3 x 6 m, 3 x 3 m and 6 x 6 m were examined however, 80.76% higher yield was recorded in 3.0 x 1.5 m spacing.

High density planting in guava has attracted attention of many workers by using growth controlling rootstocks and other canopy management practices like **Gentenprolific** guava cultivars, pruning and training system etc. More than 833 to 2700 tree/ha can be accommodated in guava under HDP, the yield vary 18-58.3 t/ha respectively (Kalra *et al.*, 1994 and Mohammed *et al.*, 1984). From CISH, Lucknow super intensive system or meadow orcharidng for guava which accommodates 5,000 trees per ha at 1.0 x 2.0 m coupled with regular topping and hedging have been found quite encouraging (Singh, 2005). Meadow orcharding or super intensive system is a modern technique of fruit production where in large number of tree ranging from 5,000-10,000 tree per hectare accommodated at 2x1m and 1x1m spacing with use of small dwarfs tree coupled with altered canopy. Due to good light penetration and distribution inside the canopy the number of well illuminated leaves increases, such conditions facilitate high rate of canopy photosynthesis which results high yield per unit area. In this case trees are required to be kept in dwarf structure by uniform height topping two months after planting for encouraging new growth below the cut end. Tree started flowering and fruiting one year after planting and yield recorded 12.5 tones/ha and increases up to 48 tones/ha at five year after planting. Canopy dimension is minimized by regular topping and hedging. Dwarf canopy size encourages air circulation and sunlight penetration into canopy centre good air circulation minimizes disease, increase temperature and reduce humidity resulted low disease occurance high fruit quality and colour development.

Tree architecture development

As guava tree grows, canopy becomes dense and inefficient with increase of tree branches. The canopy management optimizes light interception, light penetration, and utilization by the tree canopy in orchard. Canopy management involves development as well as management of tree architecture, pruning, branch bending, skirting etc. Proper control of vegetative growth is almost important in high density orcharding of guava. In case of HDP if tree do not manage proporly tree become overcrowded and shading will hamper flower and fruit production, fruit size and quality. Topping and hedging have been found valuable technique for improving fruiting pattern. Singh (2011) reported that maximum yield of 113.5 and 106.1 kg/tree was recorded as compared to 71.5 and 88.5 kg/plant under managed tree canopy when planted at 3 x 6 m and 6 x 6 m spacing enhanced light interception within tree canopies. Robinson *et al.*, 1983, Syvertsen, 1984 and Gaye (1989).

ICAR-CISH Lucknow, initiated work on tree architecture under high density planting system the results obtained is quite encouraging, Espalier tree architecture technique is as follows (Anonymous, 2019).

Espalier architecture

Espalier is a horticultural technique for controlling woody tree growth for production of fruit. In temperate climate espalier trained like wall to reflect more sunlight. The word espalier is French, which means something to rest on shoulder earlier referred only actual trellis on which a tree was trained to grow.

The iron angles post having 3-3.5m length were fixed with the help of cement and concrete at 6-7 m intervals on which 5-6 wires (tiers) were erected. First wire erected at 50-60 cm from ground level and remaining wires at 45-60 cm intervals. After erecting the structure well feathered grafted saplings were planted at 1.5x3.0 m (R-R x P-P) distance during July- August. In this system 2,222 trees per hectare were accommodated as compared to traditional system (204 plants/ha). First scaffold branches were trained at 45cm from ground level and rest scaffolds at 45-50 cm interval. Thus 8-10 main branches were allowed on 4-5 tiers, on which fruiting shoots to be promoted. In order to promote side branch notching and girdling were done as required. In this training system top most scaffolds are at 3-320 m height. The main trunk was terminated leaving 10-15 cm from last scaffolds, thus final tree height was 3.0-3.30m in espalier architecture. Total spread of tree canopy was 0.45-0.60m in both the sides (between rows).

Since the branch bending were performed on the wire erected at 90^0 angle, hence, it alter the apical dominance, resulted in better fruit bud formation at the basal portion of the scaffolds of tree. The trees trained on espalier architecture are compact and its periphery is close to the main trunk, hence, effective cropping area is more, leading to a larger production. Since most of the leaves are exposed to sunlight, hence, maximum light interception took place. High dry matter production led to more photosynthates 250-480 g per fruit) and attractive in appearance with red blushed fruit surface. Guava in this architecture bears 35 to 40 fruits (7-8 kg/tree) and projected yield 14.89 t/ha from 2nd year onward.

Advantage of espalier architecture system

1. Ease in training, pruning, fruit picking and inter-culture operations.
2. Ease in spray of growth regulators, insecticide and fungicide.
3. Minimum wastage of spray materials and more trees can be sprayed in a shorter period of time.
4. Branch, shoots as well as leaf positioning are possible in this architecture.
5. Since the scaffolds are managed in tier system hence, good air drainage helps to minimise microbial population.
6. Cost effective and easy to develop this structure.

Espalier architecture system is good option for increasing high quality guava production. The tree architecture system eases in inter-culture operations. Fewer incidences of disease and insect due to maximum exposure of the scaffolds branches. The fruits are attractive in appearance, larger in size and most of the fruits are 'A' grade size. Popularization of such architecture system will help to the farmers in production of export quality fruits and doubling the farmer's income.

Table 1: Yield attributes of guava var. Lalit trained on espalier architecture

Characters	Espalier architecture	Traditionally trained tree	Remark
Number of fruits/tree	28-35	8-15	Yield and quality attributes recorded in 2nd year after planting
Average fruit weight (g)	230-290.0	120-245	
Fruit length (mm)	61.76	55.0	
Fruit diameter (mm)	65.65	57.0	
Yield (kg/tree)	7-8	3-5.0	
TSS (^{0} Brix)	12-13.58	12.0	
Fruit surface red blushed (%)	15-25	10-15	
Percentage of 'A 'grade fruits (%)	55-61.5	22.50	Fruits diameter (85 to 90 mm), graded as 'A' grade fruits
Average yield t ha^{-1}	14.89	6.60	

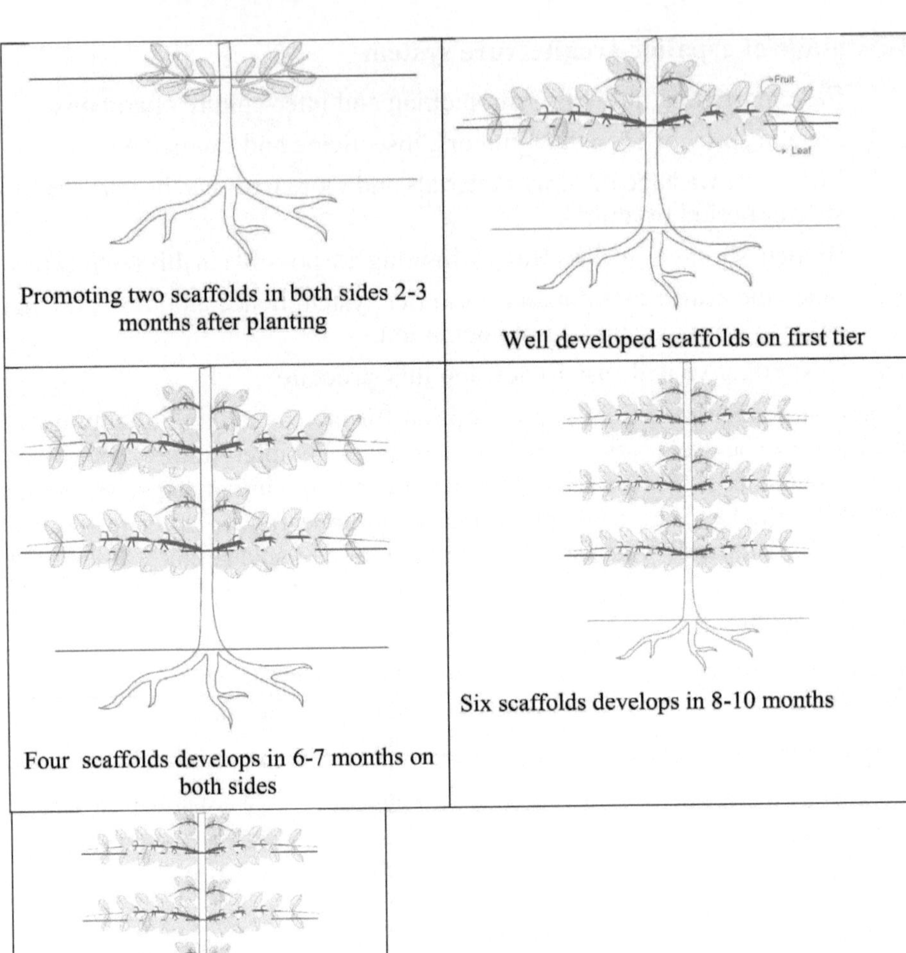

Fig. 29. Development of espalier architecture

Fig. 30. Tree laden with fruits during first year

Fig. 31. Tree laden with winter crop

All 8 scaffolds develops in 1-1.5 m on both sides

Fig. 32. Espalier architecture system

8

Canopy Development and Management of Citrus

India stands 3rd position in citrus production in world after China and Brazil. It is the most important fruit crops grown in India, however, our productivity is quite low 8.9 t/ha (Srivastava and Singh, 2006).

Production (million metric tons) 2018-19

Year	Lime/Lemon	Mandarin	Sweet orange	Others
2017-18	2.30	4.44	3.21	1.41
2018-19	3.15	5.1	3.27	1.03

During the year the volume of mandarin production was the highest among citrus fruits. In India, sweet orange (Musambi) is on 2nd 'position on production basis'. The citrus fruit availability per capita per year in India was 7 kg which was quite low, comparing world average 11.62 kg per capita per year. With increased in world population the citrus fruit availability has decreased to 7-8 kg per capita per year. It is imperative to focus on improving productivity through innovative approaches like new canopy architecture development, intensive management practices and high density planting to harness the natural resources for higher fruit production. However, not much research has been done on this aspect. Intensive orcharding and tree architecture management are complementary and high supplementary to each other. Canopy architecture development has not received due attention in evergreen crops like citrus. Small canopy is need of the hour for accommodating more plants per unit area. If canopy of citrus if canopy not attended annually they tend to overgrow and produce dense, overcrowded and unmanageable canopy and bearing in such canopy takes place on the outer periphery only.

Mandarin, sweet oranges and acid lime are three important citrus species grown commercially in India. Structure, shape, flowering, fruiting and several other physiological and horticultural characteristics of citrus plant have direct bearing with orchard management and productivity. Proper training and pruning practices helps to develop proper tree architecture in young plants, promotes photosynthesis by exposing leaves to light and air besides keeping tree in proper size and shape for effective orchard, management. Citrus orchards which have been planted at 7x7m or 6x6 m distance without proper canopy management system in such orchard overcrowding occurs after 15-20 years.

Bearing behavior of citrus

Growth in citrus takes place in three seasons viz. spring, summer and autumn in tropical and subtropical climate. The spring and summer season growth are quite important as bearing twigs develop in summer and spring. Single flowers may develop in shoot apex. The spring shoot which develop from old bearing wood are highly productive.

High density planting in citrus

High density planting is concept to accommodate more number of plants per unit area with an aim to increase land use efficency, early and high yield, with returns, high water and nutrient use. Citrus normally planted at 6x6 m distance which accommodate 277 plants per hectare while as 5x5 or 6x3 m accommodate 400 to 555 plants per ha. High yield at 6x3 m in Nagpur mandarin was recorded due to better light interception and high plant numbers, similar result was also obtained by Sharma *et al.* (1992) in 'kinnow' mandarin planted at 6x3 m spacing as compared to 7.6x7.6 m. In Punjab some farmers are growing 'Kinnow mandarin' at 6x3 and 6x4 m spacing. In Brazil IAC-5 citrus of *C. latifolia* grafted on 'Flying Dragon' rootstock and planted at 1, 1.5, 2, 2.5 and 3 m between plant and 6 m row to row, high yield was recorded in 6x1.5 m, spacing. Effect on fruit to plant quality was minor, however, fruit quality like vitamin C, TSS, juice content were better in fruits obtained from 6x3 m spacing (Machado *et al.*, 2012).

Canopy management of citrus

Canopy management in citrus is foremost important as it continue vegetative growth in tropical climate. Overcrowding and shading in canopy, reduced yield and fruit quality, fertilizer application, and application of fungicide, insecticide at targeted place and ultimately fruit picking. Suitable tree architecture, management and pruning of citrus helps to attain desired canopy model. Tribulato *et al.* (1992), reported that double row system comprise 1200 plants (2.5x2.5m x 4.5m between rows). For this kind of canopy drastic pruning needed to contain the canopy spread. In the double row planting system of canopy cumulative yield for 13 years was 493 tons per ha and 289 tons ha in 416 plants per ha. With increase in canopy volume yield also increased proportionately. High density of 800 plants per ha (4.3x2.8 m) in single row produced better yield. Double hedge row are suitable where plant canopy are more compact e.g. 'Navelina' orange.

Canopy development of young plant

Training is important aspect for successful fruit orcharding. It begins immediately after planting, tree aged 1-3 years is the best use of the limited time that pruners have. It is always better to cut small branches than large one. Diseased, dead, broken branches can be removed anytime. Branching as well as frame work must be encouraged leaving 45-60 cm above ground level, up-to this height no shoots are allowed to grow. Growers must select 4-6 well spaced branches, having 45-60^0 crotch angle in all the directions finally tree must skirted at a convenient height to avoid excessive tree canopy height.

Canopy management of bearing trees

The objective of canopy manipulation of bearing citrus orchard is to improve light penetration, canopy lightening and sap flow to the fruits and also to renew new fruiting wood. After fruit harvest, skirt trees at such height at which to reallocate flowering and fruiting to remaining part of the canopy. Skirted trees have no less fruiting it is easier to fertilize, irrigation, and spray. Light headging is done to keep walk ways free between rows, avoid cutting wood more than 10-12 mm diameter. Tree should be avoided hedged as light as possible, hard hedging is not recommended. Heavy topping should also avoid due to growth inside the canopy and subsequent loss of bearing potential.

After skirting, remove dead, diseased, damaged, broken, cross-over branches, rubbing branches water shoots, helps to open multiple windows and entry of spray materials. Sometime for better air drainage in and out of the tree, window cut pruning is done by removing interior bearing wood.

Canopy volume of citrus fruit could be measured by ultrasonic, laser and manual measurement methods. Laser measurement gives information that could be used to compute a laser canopy volume index. Laser showed better prediction of canopy volume than the ultrasonic system because of higher resolution. Automatic mapping and quantification of canopy volume of citrus trees can be determined by ultrasonic or laser sensor (Tumbo *et al.*, 2002).

Principle of pruning

Reduction of bud numbers by pruning leading to change in canopy lightening. Pruncr should only remove dead, diseased, out of place wood, water shoots. Light is critical inputs of fruit production and fruit development. Within the canopy, where light is optimal and sap flow streams are strong are of high quality, being both larger and sweet (Krajewski, 1996; Piiaway, 2000, Bevington *et al.*_2002). Among the cultural practices, only pruning and training practices can optimize light distribution and sap flow to the fruits (Krojewski and Pittaway, 2000). Pruning is the management practice most likely to optimize production and fruit size (Bevington *et al*. 2002). Pruning in citrus resulted in new vegetative growth which need to be protected from disease and pest. Pruning must not delay, as soon as the branch is seen to be out of place, train it or remove it.

Rejuvenation

As the tree advances in age, the canopy volume become large, air drainage and bearing took place on the outer peripheri of the canopy, productivity of such old plants decline. In such tree physical dimension like tree height and spread, length and tree complexity need to be reduced through rejuvenation pruning technique for making the old unproductive trees in to productive, it takes 2-3 years to rebuild canopy. For application of rejuvenation techniques the tree must be healthy and proper care and maintenance are required for the rejuvenated trees.

The severely pruned tree exhibit strong tendency of emergences of large number of shoots, 2-3 per outer growing well spaced shoots of branch should be allowed, rest be thinned out. Vigorous growing shoots tends to encourage lateral shoot growth.

The productivity can be increased by accommodating more number of trees per unit area, which is possible by efficient canopy management. High yield (33-72 ton per ha) in mandarin can be harvested from 555-5000 trees per ha. 'Grape fruit' yielded 58 tons per ha from planting density of 2020 tree/ha. 'Kinnow' on 'Troyer Citrange' rootstock at 1.83 x 1.83 m (3000 trees/ha) it starts fruiting 50-60 fruits per tree after three years (Goswami *et al.* 1993, fruits from high density planting have better in colour, juice quality and juice content. Some dwarf cultivars like 'Glodkokuy' 'Naval Orange', 'Kowanowase' and 'Anaseullis Saddreo' in mandarins have been identified for high density plantings in USSR (Awasthi and Mehta, 2000). In citrus a number of size controlling rootstocks are available viz. *Fortunella* sp, 'Rangpurlime', 'Troyer Citrange', 'Trifoliate Orange' and 'Flying Dragon'. The yield of citrus can also be enhanced by removing the upright branches and encouraging the horizontal branch orientation by pruning (Goswami *et al*., 1993). Removal of one year old shoot to half of their length is recommended to increase the yield in mandarin (Dzhincharadze and Dzhabnidze, 1989). The high density planting is practiced in 'Kinnow' mandarin by proper control of crowded vegetative growth. Light to moderate pruning improved the yield and quality marginally over heavy pruning along with alternate tree removal in eight years old 'Kinnow' mandarin on 'Jatti-Khatti' rootstock planted at 6x3 m (Sharma *et al*., 1997). Removal of alternate tree resulted good light penetration and distribution takes place inside the canopy. Yield was increased by skirt pruning at a length of 1m in 'Washington Navel Orange' (Sneath, 1988).

The genetic feature of 'Kinnow' fruit is accommodating itself in the arid zone as such it has good canopy to shade fruits and protect them from sun burning, further it can tolerate drought and bears fruit inside the well formed canopy.

9

Canopy Development and Management of Apple

Apple (*Malus domestica* Borkh) is a typical temperate fruit; more than 80% world supply is being made from Europe (Tukey, 1982). It occupies more than 85% area and 60 % production of total temperate fruits. As per 2017-18 statistics it was cultivated over an area of 30,1000 ha with 23,27,000 metric tons production (Anonymous, 2018). Austria is leading in apple productivity; however, China is a largest producer in world. Jammu and Kashmir, Himachal Predesh, Uttarakhand, are leading producer of apple in India. Baramulla of Jammu and Kashmir have recorded highest productivity (17t/ha).

Apple gives highest yield of good quality fruits in locations having long day hours with high light intensity and relatively warm days with cool nights and low relative humidity during the growing season. Flower bud initiation occurs when growth cessation of shoot takes place. Fruit bud usually develops on spurs laterally on 2-3 years old wood; however, buds may also develop on the tip of one year old shoots which are found mostly when dwarfing rootstocks are used. A good correlation was found to exist between the leaf area and flower bud formation. As no flower bud formation was found on leaf to fruit ratio below 10, in cultivar ' Granny Smith' and 'Delicious' however, 20-50 leaf to fruit ratio was found optimum for flower bud development. The sink effect of fruit is important for utilization of photosynthates produced by the leaves (Buban and Faust, 1982). As per reports available, leaf area of 190-230 cm^2 is needed for each 'Golden Delicious' fruit (Hansen,1969) and other cultivars may require 400 to 670cm^2 (25-42 leaves). With the advent of WTO regime, there are demand of high yield of good quality fruits which can only be obtained with proper balance between reproductive and vegetative pants with row, tree and branch orientation and light interception by way of balance canopy modeling. In apple canopy management had great influence on production of high quality fruits.

Canopy development is crucial for the production of high quality fruits, early and sustained yield of an orchard. Canopy development can be altered by mineral nutrition management, moisture condition, plant growth regulators, pruning and training, light interception, distribution and rootstock and scion characters. There are large number of vigour controlling scion ranging from the non branching columnar type to the spreading and non spur types.

Rootstock

Apple has wide range of rootstock which controls scion growth behavior; if scion variety grafted on M 9 rootstock, the growth of the composite tree, modified scion growth by reducing the mean length (Cannon, 1941; Rao and Berry, 1940). In addition fewer axillary buds on the primary shoot tended to grow out, hence, only few secondary shoots formed (Jasmine *et al.*, 1993; Van Hooijdank *et al.,* 2010). The scion vigor gets reduced when grafted on dwarfing rootstock in 2nd year and flowering advanced in the scion. Hence, dwarfing apple rootstock modified the scion architecture of composite tree in 1st or 2nd year.

Lockard and Schneider (1981) opined that dwarf apple rootstock control the vigour of scion by reducing the basipetal movement of Indole-3 acetic acid (IAA) from scion to root that limit root growth or cytokine in biosynthesis and consequently the amount of root produced cytokine supplied to the scion in the xylem reduced. Graft union of the composite tree (rootstock and scion tree) may contribute to rootstock induced scion dwarfing by limiting the supply of water, nutrients and hormones into sciòn (Atkinson *et al.,* 2003). A dwarfing apple rootstock supplied lower concentration of cytokine to the scion, due to which limited shoot extension takes place (Jones, 1973; Kamboj *et al.,*1999).

Crotch angle

The composite tree with wider crotch angle is good from branch strength point of view. It is said that wider the crotch angle stronger the branch and narrower the crotch angle weaker the branch. In central leader training system, wider scaffold branches are foremost important. According to Tasked and Shoemaker (1972) wide crotch angle form strong unions between branches and adjacent tissue of trunk and bear heavy crop load. The tree with narrow crotch angle are weak and prone to tear down due to bark inclusion between branch and tree trunk, further crotches may facilitate the entry of pathogens and insects.

Spur type apple cultivars have more upright branch growth as compared to non-spur type. Lower branches have wider crotch angle. Ottawa 8, rootstock had tendency to produce primary branches with wider crotch angles than semi dwarf to standard rootstock (Warner, 1991).

The rootstocks besides affecting growth and yield of scion cultivars, it also influenced the photosynthetic efficiency and fruit quality (Isarva, *et al.*, 1983; Larson *et al.*, 1985 and 1986). If a scion variety raised on dwarfing rootstock its major parts of the foliage exposed to sunlight, thus helps in increased high quality fruit production. Most of the Delicious apples are having upright growth habit produce quite a high proportion of poor quality fruits in the inner side of the canopy. Recent trends of spreading the branches led to harvest maximum light energy. Further, bending down the branches improve the photosynthetic efficiency. (Blinovski, 1970), the fruit quality the result of better light interception. Sharma and Chauhan (1992) found best quality fruits in Starking Delicious when grafted on M 7 rootstock with spindle bush canopy form, than modified central leader system.

Tree architecture

Tree architecture is the potential tools to manage the canopy so that maximum possible solar radiation could be harnessed which is a limiting factor in the hilly region. With the development of dwarfing rootstock and spur type cultivars in apple, several training systems are in practice. Open tree canopy have well exposed canopy area which promote high fruit numbers, good fruit color and fruits are larger in size (Wunsche and Lakso, 2000). The open canopy form can be achieved by removing the central leader after 1m height of ground level. The centre of canopy remains open, permitting entry of adequate sunlight throughout the tree.

Lakso *et al.* (1989) described the modification of the central leader tree form named the Palmette leader designed to improve light penetration. This type of canopy permits maximum exposure of branches to sunlight.

Wertheim (1968) suggested that slender spindle tree forms have narrow canopy with dwarf tree to allow high tree density in single, double, triple or multi-row beds to improve early yields in apple orchard. The width of the canopy is less than 2m which ensure branches be exposed in sunlight. Slender spindle has been found suitable for cultivar, like 'Golden Delicious' and James Grieve and other relatively weak growing ones.

Chalmers and Vanden Eude from Australia developed Tatura Trellis architecture, aimed to develop continuous canopy of fruiting wood half meter deep to maximize yield at the same time to facilitate mechanized management of orchard. Two main branches are allowed to develop at 50 to 60^0 crotch angle, it require summer pruning to regulate canopy depth and shading.

Lincoln canopy was developed in 1970 by Dunn and Stolp, in this method all branches are trained to set of six wires on each side of the tree. All branches are trained in a single horizontal plane. The small tree canopy offers easy to mechanical spraying, pruning and harvesting. It forms a continuous canopy of about 3m wide which resulted maximum light interception and penetration.

Carbonneau and Lespinasse (1989) found double vertical axis, MIA 30 (1 or 2 inclined axis) and double inclined axis similar to the Tatura Trellis and A,V, were found best for total exposed of leaf area and potential photosynthesis point of view.

Pruning

The un-pruned tree look shabby and bears small fruits of poor quality due to low interception and distribution of sunlight in the inner parts of the canopy. Competition between vegetative and reproductive organs, is particularly severe during flowering in following 3-4 weeks, pruning during this period can increase both flowering and fruit set. Removing the leader and inner main branches and by shortening the laterals on the remaining scaffold branches to 25 to 30 cm. Light intensity inside the canopy was increased by 2-3 times and the yield was doubled in apple (Groza, 1972).

Summer pruning has been found effective to increase light intensity and red coloration of fruits. Marini and Barden (1982) observed that summer pruning has influenced on the slowing of the rate of starch disappearance from fruit flesh, pre-harvest drop, severity of water core and bitter pit were suppressed in 'Golden Delicious' and 'Stayman' apple cultivars.

Intensive Orcharding (High density planting)

High density allows greater early productivity, return and sustained high yield with good quality fruit (Wertheim *et al.*, 2001). In comparison to large tree, dwarf trees have greater portion of canopy is illuminated and hence, such canopy require lower pesticide inputs and higher labor and water use efficiency. Early high productivity of intensive orcharding system is partly based on their greater leaf area per unit of land area, resulting increased light interception of photo synthetically active radiation (PAR) as compared to low density plantation (Jackson, 1989). Full field and multi row system can intercept more light than single rows, single row has added advantage of easy management and machinery movement, with increased in tree density the tree size decreased. But leaf area and light interception increased (Wagonmakers,1991b; Wertheim *et al.*, 1986).

Hampson *et al.* (2004) reported that reduction in trunk cross sectional area (TCSA) were evident as early as 2nd year when tree density exceeded 3,000 trees/hectare; he further reported that a density of 1125 trees per hectare on M-9 rootstock found too less, trees in such density never filled the allotted space and much light fell unused on to the alleys. Lakso and Robinson (1997) reported that if tree received 50-70% light falling on the tree canopy is considered optimal.

Light interception by the orchard is directly affected by planting density. Maximum light interception by planting trees to closure may reduce fruit quality. Hence, an optimum density is needed to balance the yield and quality. Early high production in high density orchard system is due to the rapid development of leaf canopy in the first few years after planting. High leaf area early in the life of the orchard is achieved by planting at high tree densities, by planting large, well-branched (feathered raplings) and by minimal pruning (Barritt *et al.*, 1991). Further, Barritt (1989) found that tree density was more important than training system or rootstock for improving light interception and yield of 'Granny Smith' apple, therefore, the bed system planting was standardized to achieve high light interception, with more even distribution of foliage. Palmer (1988) reported that bed system can intercept up to 80% of photosynthetic active radiation from the end of June until October and yielded 78 tons per ha in the 3rd year.

Fig. 33. Apple on espalier architecture

Fig. 34. Spur type

Fig. 35. Apple variety Red Spur

Meadow orcharding

Super high density planting was first developed in 1969 from England and the first experimental orchard was planted at East Malling Research Station, with density of 20,000 tree per ha on dwarfing rootstock and Long Ashton Experiment Station, planted 70,000 trees per ha which was called meadow orchards using ultra dwarf rootstocks (M. 27) and sprayed with succinic acid; 2, 2-dimethy hydrazide to control tree vigor. In super high density orcharding bearing starts from 2nd year and gave 3-6 tons per ha production and 40-50 tons per ha yield obtained in the fourth year after planting. However, sprays with SADH and low light penetration inside the crown had negative effect on fruit quality. Therefore, it was suggested that super high density orchards should not have more than 7300 trees per ha in 5-6 row beds with 3m paths between the beds on M.26 rootstock coupled with summer pruning and fruit thinning up to 20-30 fruits per tree.

In some apple cultivars, excessive and upright vegetative growth and low fruitfulness may be experienced. This leads to difficulty in pruning, training and unsatisfactory tree structure, delayed cropping and high cost of management. Orientation of apple shoots and seedlings towards horizontal position has been reported to reduce terminal shoot growth (Myers and Ferree, 1983; Tromp, 1986) and enhance spur formation (Mullins, 1965; Tromp, 1986). Shoot orientation by increasing the shoot to the horizontal position is known to increase the number of lateral flower production in apple (Wareing, 1970). Sharma and Jindal (1992) reported that horizontal plantations retarded linear shoot growth, internodal length and radial growth of shoot and trunk. Retardation in growth was proportional to the degree of tree bending from vertical position and thus 30^0 plantation registered maximum reduction in canopy volume. Production of maximum fruit spur was found in 45^0 and 30^0 plantation in 'Royal Delicious' apple.

Fig. 36. Apple on single axis training system

Fig. 37. Apple var Starking Delicious

Fig. 38. Apple var Red Spur in Y shape architecture

Canopy development and management practices are being followed in European countries are advanced, however, most of the apple orchards in India are still on local seedling rootstock which are vigorous in nature. The trees on seedling rootstocks are large and difficult to manage. Further the high density orchards are being encouraged on growth controlling rootstock, in which canopy can be managed. The smaller and well managed canopy under high density planting system could intercept more solar radiation resulting high total yield of quality fruits.

10

Canopy Development and Management of Pear

The common pear (*Pyrus communis*) is known as French Pear, Asian pear also called Chinese pears, Japanese pears, Oriental pears, sand pear, salad pears and apple and pear are group of pome fruits derived from *Pyrus ussuriensis* and *Pyrus serotina. Pyrus communis* is susceptible to fireblight disease, however, *Pyrus pyrifolia* is resistant to this disease. Pear is only temperate fruit grown on the hills to southern parts of India. Jammu and Kashmir is leading in quality pear production.

The pear tree is huge in size there is natural progression from a single-shoot tree toward the long slender and upright branches which pose difficulty in spray, pruning and fruit-harvest. Further poor light distribution throughout the canopy, low early light interception, leaf area index and fraction of land covered by the canopy leading to poor fruit set and quality. These disadvantages have resulted in wide-spread efforts to reduce tree size, increase tree density, accelerate canopy and yield development and to improve canopy geometry to overcome the limitations of large tree. Ito *et al.* (2004) reported shoot reorientation from vertical to horizontal position increased carbohydrate and related enzyme activities in the bud and shoot internodes of 'Kosui', Japanese pear, which accelerate flower bud formation. Canopy management deserves greater attention by exploiting the rootstock and scion combinations, training or tree architecture, pruning and cultural practices as well as use of growth retarding hormones.

Central leader tree architecture

This is the most common type of tree architecture, called central leader training system. Mostly this training system adopted in region where solar light is not a limiting factor. It is characterized by one main leader or vertical trunk and several whorls of scaffolds branches, at 60-75 cm above ground. The scaffold branches get progressively shorter from base to top. This allows maximum light penetration into the canopy. Each group of scaffolds branches have 45-60 cm gap. Each whorl consists of 3-4 well spaced branches. Tree architecture development approach is to provide fewer large, permanent opening for light penetration into canopies restricted into geometric form. Thin restricted planes of foliage such as narrow hedge rows, tree walls, and A, V. or T forms tree. This approach generally requires severe geometric restrictions of the canopy, expensive support structures, and significant labor to place and maintain the branches in specific locations. The value of these different tree forms lies in their light distribution properties and the ultimate improvement in fruit yield and quality. Pear cultivar 'Spadona' on 'Quince A' rootstocks trained as oblique, hedge, slender spindles, form 4 branches trained at 45^0 angle and many secondary branches. High yield was obtained by training as hedge, after two years, the irregular palmette gave the highest yields in comparison to others. Pear cultivar 'Conference' was trained on four system of trainings viz. angled trellis, 'slender spindle', 'vertical axis' or 'Y trellis'. The tree trained on 'Y' trellis had the greatest spread after 5 years and the vertical axis and 'slender spindle' trees were the tallest. The 'Y trellis' trees have significantly higher yield than other training system (Kappel and Brownlee, 2001).

The benefits of smaller trees led to the development of the central leader tree forms by Heinicke (1975) and Mckenzie (1972). The system has a pyramid shaped tree with tiers of branches spaced along the trunk. Years of experience led growers to increase the distance between the bottom and second tier to a minimum of 1m to have gaps in the canopy to increase the light penetration to the bottom tier.

The single, double or multiple row planting system leads to improvement the early yield. The percent light interception by a single-row planting at 2000 trees/ha was about 10% less than for one at 2667 trees/ha or for a 3-row planting at 3,570 tree/ha in pear cultivar Doyenne du Comice. The trees on slender spindle bushes or super spindle bushes have not been found economical in pear (Teeffelen and Teeffelen, 1993).

Tree Forms: Planar canopy have been developed to overcome the problems of light penetration into thin canopies. The thin narrow hedgerows, tree walls, vertical low and high trellis, ebro trellis and 'A', 'V' and 'T' trellis since the foliage and limbs are restricted into a single plane. These tree forms usually have a dense canopy that is essentially non-transmitting in ebro trellis, and 'T' type horizontal planar canopies the light levels decrease drastically from the top to bottom. In 'T' trellis (Lincoln canopy) tree experience high light transmission through the canopy. Inclined 'V'-or 'A'-shaped canopies were also developed for mechanical harvest and very high yield but they overcome many of the horticultural problems of the horizontal canopies. Light distribution in 'A' and 'V'-shaped canopy depend on light transmission through the canopy and direct light exposure of the underside of the trellis through the open gap at the top of the canopy arms. Reinhoudt (1997) observed early yield per tree, when the pear cultivar 'Conference' planted at 3000 trees/ha and trained on the 'V'-hedge system with 3-branches' yields were highest with trees planted in double rows at 12,000 trees per ha, trained on 'V'-hedge system.

Bist and Sharma (1992), showed that paclobutrazol and promalin increased number of vegetative and flowerings shoots, canopy spread and yield of treated trees. These chemicals reduce the tree height and induce spreading of the branches with high productivity in 'Gola' pear. The growth retarding chemicals are helpful in tree canopy development and shaping.

Multi-row and bed system

To achieve the high light interception with even light distribution, Palmer and Jackson (1997) developed a bed system with a more even distribution of foliage over the orchard floor. In case of cultivar 'Doyenne du Comice' yield/ ha increased with planting density from 2,000 to 4,000 trees/ha. However, leaf area per tree decreased linearly with density. Three-row and 5-row beds tended to have higher leaf areas than single rows at a given density trees in 3-row and 4-row beds tended to transmit less light than trees in single rows (Wagenmakers, 1989). Further, Pepelyankov and Grnevski (1986) found closest spacing have limited growth in width, encouraged growth in height, gave the lowest yield per tree but highest yield per unit area and reduced fruit quality, spacing at 3.5 x 1.1m was most suitable for pear on 'Quince M A' rootstock. Pear planting under high density gave higher yield during productive years, but as the time advances fruit size reduces. Yield increased with increasing trunk cross sectional area and percentage canopy occupation (Kim *et al.*, 1987) was going to deteriorate (Assaf, 1998).

The density cum system of planting had remarkable effect on vegetative growth. 'Bagugosh' grafted on *Pyrus pashia* rootstock, the plant under high density exhibits excessive and vigor vegetative growth, which could be effectively be controlled by paclobutrazol application (10 ml per lit), which resulted good light penetration in the inner canopy of densely planted system (Gupta and Bist, 2005). The paclobutrazol greatly reduced the increment in canopy spread, similarly system of planting had influenced the canopy spread increment, too close planting tended to increase the plant height due to very little space are available for plant spread.

Pruning

Annual pruning is required for its proper growth, flowering and fruiting. Winter pruning of large tree is designed to ensure adequate light penetration. However, reports on application of summer pruning are not available in pear.

Pear need regular winter pruning (dormant pruning to encourage good fruiting and renewal of excessive spurs were also carried out.

It has been observed that the growth pattern of pear cultivars suited best to central leader tree architecture as it has strong apical dominance. However, it responded well to advance tree architecture development and management. The tree bears heavy fruit load due to which the branch breaks occurs. However, the traing on trellis system proved good for the pear.

11

Canopy Development and Management of Peach

Peach (*Prunus persica* L. Batsch) is a small to medium sized open topped branches, medium stocky, spreading as well as upright growth habit. The Peach is the native of (china, currently around 3000 peach cultivars are available in the world over (Faust and Timon, 1995). It is mainly grown in sub-mountainous regions of Jammu and Kashmir, Himachal Pradesh, Punjab, Uttarakhand, Nilgiri Hills, Jharkhand and north-eastern states. Low chilling peach cultivation is gaining momentum in northern plains i.e. Punjab, Haryana, Delhi, Uttar Pradesh, Jharkhand, Madhya Pradesh and Chattishgarh. As per 2017-18 peach was cultivated over an area of 19,000 hectares with total production of 117 thousand metric tons (Anonymous, 2018).

The peach is highly precocious bearer, amount of sunshine exposure is important factor influencing flower bud initiation and fruit quality. In general plant dry matter production is proportional to light interception (Monteith, 1977; Russell *et al.*, 1989). It bears fruit on shoot developed in the previous years from terminal and axillary vegetative bud. The exposure of canopy to maximum light interception is essential for exploitation of genetic constituents of the cultivars. Canopy development and management is essential to obtain high yields of good quality fruits. Unmanaged canopy resulted in large number of small size fruits with poor coloration.

Tree architecture development and management

A balanced tree architecture developed in peach to give the shape or build strong framework of the tree in order to support maximum good quality fruits. The training aim of is to utilize the available space and sunlight to the maximum extent for production of quality fruits. The peach canopy develop according to the agro-climatic condition. In India, where sunlight is not a limiting factors, modified central architecture preferred. On the other hand where it is a limiting, vase or open centre canopy advocated. Conventional, vase and central leader peach trees generally grow faster and carry most of their fruits ripens at the earliest. This decreases productivity and taxes unnecessary management costs. The aim of training is to minimize such effects and optimum and uniform light interception within the canopy. The trees are trained to increased uniformity of light interception and distribution within the canopy and lessen the shadding effect. In Australia different tree architecture were compared, in 'Redhaven' peach, cumulative yield of 6 years old was highest in Palmette (149 tons per ha) followed by 'Tatura trellis' (143 tons per ha), Lincoln canopy (119 tons per ha), Central leader (109 tons per ha) and 'Vase' canopy (86 tons per ha). However, the color development was most uniform in palmette, whereas in other systems, there was reduction in color development from the top to the bottom of the canopy.

Open centre canopy architecture

Open centre canopy architecture has become very popular due to high yield efficiency. The central leader is removed 1m above ground level and 4-5 well-spaced scaffold branches are retained; both primary and secondary scaffold branches produces fruit-bearing laterals. This kind of tree architecture keeps the centre of the tree open permitting entry of adequate sunlight in whole canopy layers.

'V'-shaped architecture

This tree form 'V-shaped' canopy architecture or 'Tatura Trellis' canopy is becoming popular in peach due to high yield- efficiency. 'V' shape or perpendicular (KAC-V) system is another widely used high density system for peach (De Jong et al. 1994). This system has two primary scaffolds oriented perpendicular to the tree row. Summer pruning and topping are employed to control vigours growth of the tree. In this system less pruning is done as compared to cordon system. In this tree forms two limbs of each tree are trained across the inter-row spacing at 60^0 angle from the horizontal, forming a 'V'-Shaped canopy. The canopy is supported by permanent trellis. Secondary scaffold branches develop along with each primary limb forming fruiting canopy. Gross man and Jong (1998) examined peach in four system of tree architecture viz perpendicular 'V'(KAC-V) system high density perpendicular (H1D KAC-V) system, a cordon and an open vase system compared to the 'V' system. He reported that on the ground basis, the HiD KAC. V shape system had the highest crop yield and the open vase system had the lowest. The 'cordon' and 'HiD system' intercepted more light and produced more fruiting stem, and leaf boomers was lower in the cordon system than in other system.

Light environment in peach tree canopy appears determinant, as shade decrease the fruit growth increased (Chartzocelakis *et al.*, 1993; Floored and Lakso, 1989', Tustin *et al.*, 1992). Cultural practices which limits shade and promote lateral shoot growth in the canopy may improve fruit quality.

High density vase canopy

High density vase canopy developed like a standard orchard except that the scaffolds are tipped in the winter to stiffen them and the trees are summer sheared to limit the tree size when they had occupied allotted space. Two scaffold vase are new tree forms where trees are trained with an open centre and a single scaffold is allowed to develop in each direction. This becomes a flat tree about 1.2 m wide and 3m hight but with different scaffold structure. Belgian Fence architecture are raised on 1.2m spacing at 45^0 angle in planting hole with top inclined in the row. Vertical upright shoots are allowed to develop from the inclined trunk which summer pruned to from a flat wall of fruiting surface of 4 feet thick. Another form of Belgian Fence tree forms are planting 2 tree in single hole inclined at 45^0 angle in the opposite direction at 2.4m distance. This system offers advantage for early fruiting but required support to hold the system in position. Sansavini *et al.* (1984) planted peach at, 5 x 3m and trained to the 'Early Palmette', 'Delayed Vase' and 'Free-Central systems' and were pruned uniformly, showed high yields in the Delayed Vase and Free-Shape training system than palmette and hedgerow systems.

Desalvador and Dejong (1989) compared light interception and penetration into canopy of different pruning configurations and found that mean light interception was similar in 'Y' shape (74%), 'open vase' (71%) and 'central spindle' (89%) architecture. However, sunlight distribution into the 'Y' shape canopy was 35% higher than the 'central spindle canopy' form.

Branch angle

The aim of tree architecture is to minimize shading effect inside the canopy. Shoot growth, flowering and fruiting depend on branch angle and position of node which are duly controlled by endogenous are factors rather than environmental fact Dann *et al.* (1990) noted that fruit near the tip mature earlier on the lower angled trees. As angle to the horizontal increased between 15^0 and 90^0, the gradients were negative for flower density and fruit growth. The fruit weight and volume decreases linearly as the angle become more vertical. The flower density increased on shoots towards the tip as the nod angle become more horizontal. In peach maximum fruit weight was obtained on Crotch angle between 15^0 and 45^0.

Pruning

Peach bears fruit laterally on the previous season's growth and once a shoot has fruited it never bears again in life. This is the main reason of tip-bearing habit. The objective of peach pruning is to reduce barren and unproductive parts and facilating light penetration for excellent fruit quality and color development. Dormant pruning of large tree is designed to ensure adequate light penetration. Summer pruning increased the light intensity in the cropping zone. Improvement in light penetration following summer pruning was reported by Miller (1982) in peaches. Blake (1916, 1917) initiated one of the first summer pruning experiments with peach in the United States in 1912. Summer pruning enhanced the color improvement but some researcher reported sun-burn in fruits. Severe pruning contributes to an increase in young shoot growth as it limit shade and improve the nutrient supply to shoots by the roots. (Cannell and Kimeu, 1985). Fruit weight and soluble solid concentration decreased, when fruit load increased. Siham *et al.* (2005) worked on the fruit and shoot number management in peaches which is important consideration to improve fruit quality. He further reported that high pruning intensity enhanced shoot growth, the increase in vegetative growth may explain increased in fruit growth as it enhanced available photosynthates.

Vegetative growth increased with increase in pruning intensity due to optimization of light environment inside tree likely to promote photosynthesis rates. Less number of shoot bearing fruit with severe pruning could improve the distribution of available mineral elements within aerial parts of the tree (Cannell and Kimeu, 1985).

Peach growers were used summer pruning as a danagement tools to help, reduce excessive tree growth and shading. The summer pruning affects light penetration within the tree, return bloom and fruit-set in the season following treatment. The summer pruning also influenced the canopy environment and fruit color intensify. Gerdts (1987), Day *et al.* (1989) reported that pre-harvest and post-harvest peach tree pruning by hand-topping, interior water sprout removal and their combinations resulted in increased photosynthetic photon flux densities (PPFD) within canopy. Pre-harvest summer pruning in the following years on the same tree 23 and 28 days before harvest, increased fruit size, weight and redness than to un-pruned.

Peach respond well for tree architecture management as well as development by training and pruning practices. It also gives good yield under high density planting system. Different plant geometry has been followed in peach which have encouraging response.

12

Canopy Development and Management of Cherries

Sweet cherry (*Prunus avium* L.) occupies an important position among temperate fruits. It is a tall, attaining a height up to 18m or above and pyramidal form. It is an interesting and popular fruits of temperate climate. The branches are usually erect and leaves are large and thin, the flowers borne on one-year-old spur and shoots. The spur elongates in almost a straight line for several successive years because the terminal bud is leaf bud. Sweet cherry fruits are mostly consumed fresh and also used for processing purpose, syrup and some type of alcoholic drinks etc. It is also rich in antioxidants and contains compound which helps in pain relief of arthritis, gout and headache (Serrano *et al.*, 2005; Usenik *et al.*, 2010).

Siberia is largest producer of cherry which accounts more than 7% of world's produce, followed by Russia, Turkey, Ukraine, Poland, USA and Iran. It is also cultivated in India mainly temperate region of Jammu and Kashmir, Himachal Pradesh and Uttarakhand. Jammu and Kashmir is a leading producer of sweet cherry. Over the last two decades demand of quality fruits from all over the world has been increased resulted in development of new size controlling rootstocks, cultivars specific to agro-climatic zone, method of tree architecture management suitable canopy shape and pruning systems for optimum yield of high quality fruits.

Rootstock for canopy architecture

For successful cultivation of sweets cherry, good rdootstock is most important factor. Right kind of rootstock not only influence graft quality, but also long term efficiency of the orchard. Rootstock influenced the tree growth, yield performance and fruit quality. Mazzard rootstock induced higher tree vigour, higher yield performance gross profit, where as colt induced high fruit weight, fruit dimensions, size and shape, stone weight on 7 cherry rootstock at Serbia (Milosevic, *et al.*, 2014). 'Stark Hardy 'Giant' (SHG) cultivars had the lowest tree vigour on 'Mazzard' and 'Colt' rootstock. High yield efficiency of Sunburst variety on 'Mezzard' rootstock and 'SHG' on 'Colt' rootstock. (Milosevic *et al.,* 2014). He further reported that 'Colt' rootstock when grafted with 'Stark

Hardy Giant', 'Sunburst' 'June Early' were found most surtable for growing under high density planting system.

Central leader tree architecture

The central leader retained and all other upright growing branches were removed by cutting close to the trunk only 3-5 wide angled frame work branches at 20-25 cm apart, spirally arranged around trunk. The lower most branches were allowed to grow at 40-60cm above the ground. The tree is susceptible to crotch splitting, hence, strong scaffold is required to develop, scaffold may be headed back when they grow long and un-branched.

Tarura trellis architecture

Tatura trellis system has been found promising on dwarfing rootstock. When 'Bing' and Van cultivars grown on *Prunus mahaleb* rootstock at planting density of 2222 and 3175/ha, after 3 years, yield obtained 31-39 and 9-10 t/ha respectively. It is necessary to chalk out strategy for improving canopy and source-sink relation ship which is important for high fruit quality. Many training systems viz 'Steep leader', Vogel central leader, Spanish bush, Kym green bush, upright fruiting offshoots tall spindle axe, super spindle axe have been evaluated for higher yield and efficient harvest (Zahn, 1992, Robinson, 2005). Above training systems provide early cropping and high yield and enhance cropping efficiency and fruit quality by enhancing the influx and light distribution (Lauri and Claverie, 2005).

Usenik *et al.* (2010) studied the leaf to fruit ratio of six years old 'Lapin's' cherry on 'Gisela 5' rootstock and leaf- fruit ratio was manipulated 38 days after full bloom of fruit development stage. High leaf area to fruit ratio influenced significantly darkest fruit color, higher fruit mass, higher total soluble solids contents and higher ratio between sugars and acids which corresponded to better ripening stage. While low leaf to fruit ratio influenced prolonged ripening process and delayed fruit maturity of 'Lapins' variety of sweet cherry.

Whiting and Lang (2004) reported that fruit quality decline in 200 cm^2 canopy leaf area in cherry. A very intensive cherry orchard with trees trained on a vertical crown canopy. Parnia *et al.* (1986) found that high and very high density orchards of sour cherry had limited trees height of 2.5-3.0 m and the hedge row width 1.5m where as very high density was trained on vertical cordon obtained the highest yield than 'palmette' trained.

Pruning

Pruning in bearing tree is done to remove weak and unnecessary branches to keep the centre of the tree open and facilitate the formation of new shoots. Cherry required less renewal pruning as compared to other deciduous sour cherry and other fruits. In sour cherry, annual shoot growth should be 10-20cm long in fully grown bearing trees for maintenance of yield. Fruit buds are borne on lateral shoot of one-year-old terminal growth, while fruiting spurs are formed on 2 to 3 years-old vigours shoots and 4-5 years old less vigour shoots. A single crop of fruit weakens the spur and they are short lived, hence most of the crops are borne on periphery of the tree, pruning should include removal of long branches from origin point. In cheery there are also problems of growing long un-branched shoots hence, proper pruning is essential to optimize the growth by increasing the light interception within the canopy. In case of upright growing cultivars i.e. 'Black Tartarian', removal of upright growth and heading back of out ward growing shoots, caused spread of shoots. In case of 'Early River' cultivars which produce long whippy leaders may need some tipping. Summer tipping of shoots in 'Van' and 'Merton Glory' consistently reduced mean shoot length and tree size and floral buds increased (Webster and Shepherd, 1984).

Most of the sweet cherry cultivars are having spreading to upright growing habit. For high density planting system, up right growing cultivars are suitable. Sweet cherry is raised on sour cherry rootstocks, which are vigorous in nature, hence, mostly orchards are on traditional spacing. However, with the introduction of clonal type cherry rootstocks in India, HDP orcharding has gained momentum.

13

Canopy Development and Management of Plum

Japanese plum (*Prunus salicina*) is usually eaten as fresh while as European plum (*Prunus domestica*) required to process before consumption. Plum ranks next to peach as for as its economic importance is concerned. It is a delicious juicy fruit, prized both for its exquisite fresh fruit flavor and in processing industry. Due to its typical chilling requirement it is grown in cooler region. Two types of plum i.e. Japanese plum and its hybrids and European plum are grown in India. Plum bloom early in season and hence, are prone to spring frost injury. Selection of relatively higher orchard site with good air drainage is suitable for plum cultivation. A gentle slope is desirable rather than enclosed plain area, this would ensure adequate air and water drainage. Plum is cultivated over an area of 23,000 hectare with production of 89,000MT (Anonymous, 2018). Jammu & Kashmir, Himachal Pradesh, Uttarakhand are leading producer of plum in India.

Tree architecture

Plum trees are mostly trained on open central leader and 'modified central leader system. Japanese plum is more adaptable to the open central leader architecture. In modified leader training, the central leader dominates over the scaffold branches. Primary scaffold branches should allow at the interval of 15 to 20 cm apart along the trunk. Secondary branches are to be selected during the 2nd, 3rd and 4th dormant seasons. At the end of 4th years of growth and pruning, 7-9 well spaced secondary branches are obtained, which do not hinder light penetration into the centre of the tree and thereby stimulate fruiting bud.

Under high density planting system, low vigour rootstocks, slender spindle or spindle bush tree architecture were found appropriate (Grzyb and Rozpara, 1998; Hrotko *et al*., 1998; Meland 2001; Cmelik *et al.,* 2002; Gaivrilescu *et al*., 2004).

Major objectives at tree architecture is to obtained high production with high fruit quality and to improve access. Studies have shown that the effect of tree architecture (Training) methods and tree density on root distribution light interception, leaf area index and water relationship of the plant. Training methods also influenced tree structure and hence, the microclimate and some aspect of water relationship. Tree trained on 'Lincoln', 'Vase, Palmette' and 'Tatura trellis' systems, the 'Tatura trellis' has the greatest yield (50-60 t/ha) followed by the 'Palmette', 'vase' and 'Lincoln canopy' (Chootummatat *et al.*, 1990). Tree trained on Tatura trellis used more water than the other systems of training but the difference was between 9-12%. Trees with high crop load used slightly more water than those with a low crop load. The difference in canopy configuration, the water use were not large, but yield differences were found considerable, 34 tons per ha for the Vase to 62 tons ha in 'Tatura trellis' architecture, (Chootummatat *et al.*, 1990). 'Tatura trellis' and 'Palmette canopy' architect used more water than the 'Lincoln canopy', this was associated with their ability to extract water from the deeper layer of the soil.

Pruning

Fruits are borne laterally on short spurs in plum, as the tree advances in age the bearing area become dense hence, the older trees will need thinning out pruning to facilitate spraying and allow better light penetration. In bearing trees, pruning is directed towards the breakage of branches due to heavy cropping and promoting continuous new growth. Usually japanese plum pruned more heavily than European plum. This is however, not universally true for all Japanese plum cultivars, 'Santa Rosa' being a such exception (Christopher, 1979). However, heavy heading back of the shoots should be avoided as it would only encourage development of long upright shoot and too dense top, only light thinning should be adopted in order to admit sunlight, facilitate spraying operation and to develop a healthy spur system (Tasked and shoemaker, 1972). However, in case of 'Santa Rosa' plum, 75 per cent heading back resulted highest trunk girth, shoot length, fruit set, highest fruit length, breadth and weight no heading back results poor fruit quality with high yield (Kumar *et al.*, 1992).

High density planting system

High density planting are being advocated in plum to accommodate more number of trees per unit area. Koroid (1987) has reported yield increased up to 168% by reducing the planting spacing to 6.2m and training trees on 'Tatura trellis'. 'St. Julian K' is a true dwarfing rootstock for plum which has been developed at the East Malling Research Station (Webster, 1981) and 'Pershore' rootstock for semi-dwarf type tree. These clonal rootstocks are helpful to keep the tree canopy in control and efficient exploitation of solar radiation.

14

Canopy Development and Management of Kiwifruit

There are number of eating type kiwi fruits in world but *Actinidia deliciosa* Chev, is most common. It was the recently introduced fruit in India. It was origin of China the oval fruits have flavor of gooseberry. It is egg sized, fuzzy, brown skinned green fleshed kiwifruit. Fruit is a berry, where as *Actinidia chinenses* a yellow fleshed fruits. It holds a great promise for commercial cultivation in low and mid hills of entire Himalayan region. It is predominantly grown in Arunachal Pradesh, Jammu and Kashmir, Himachal Pradesh, Sikkim and other similar climatic zones. Kiwifruit is a precocious and prolific bearer vine. It is unique in many ways; while most other fruits are attractive in the appearance, it is dull brown in color similar to sapota, but the cross section of fruit is surprisingly very beautiful. Kiwifruit is dioecious in nature, long growing unbranched shoots, with annual growth of 3-5m each year, require topping in summer to keep plant in shape. Fruits are borne only on current season's growth arising from the bud developed in previous year, 3-5 basal buds of current growth are only productive. Kiwifruit vine, like grape require structural support for commercial production. The vegetative growth and fruit load in these plants are heavier than grapes necessitating the erection of stronger supporting structure and heavier pruning and shoot positioning. In vigours growing vine canopy development and management is a very important operation. The canopy may be manipulated through various means, for efficient utilization of solar radiation.

Canopy architecture development

The prime objective of architecture development is to establish and maintain a well-formed framework of main branches and fruiting arms in order to keep balance between vegetative growth and fruit production throughout the life span of the vine. Training facilitates pruning, spraying and harvesting operations. Training of vines begins with the planting and continues throughout its life. Since vegetative growth and fruit load in kiwifruit are heavier than grapes, they require stronger supporting structure and heavier pruning to produce quality fruits with regular yield every year. Kniffen and Pergola systems are commonly followed for free architecture development. A flat topped network or criss-cross wire is prepared to train vines in pergola or bower system, for the Kiwifruit, this is the best training system however, it involve more cost to erect the system. In this system the vines are well exposed to sunlight, hence, micro-climate is improved.

Bower or pergola tree architecture

This tree architecture usually suitable for vigours growing varieties e.g. 'Hayward'. The kiwifruit vines are trained on the network of the GI wires fixed at 20-30 cm intervals in criss- cross manners. The GI wire network are supported by angle Iron of 7-8 feet length.

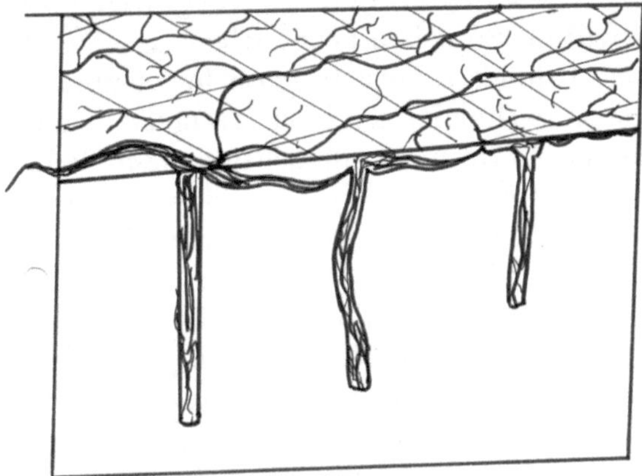

Fig. 39: Bower or pergola architecture

Kniffen architecture

A single wire fence is commonly adopted through another wire form the kniffen system at 2m hight above the ground on concrete pillars. Kniffen architecture used for the low vigour growing vines. In this architecture 4-5 wires are erected at 45-60 cm intervals from ground level. The scaffolds branches on both the directions are, on which fruiting takes place. Kinffen architecture is less expensive than spreads pergola (Grape Kniffen).

Telephone type tree architecture

Telephone architecture usually adopted in variety which are less vigours in nature. Main leaders are allowed to grow up to some height in single stem after that it is headed back and 3 sub branches or primary scaffolds are encouraged.

Another form of tree architecture is overhead trellis, the vine is headed back at the height of about 30cm above a bud just after planting to promote vigorous growth. The main shoots are allowed to grow in the first few years to develop in a strong trunk. The main shoot is tipped to promote side arms; laterals are also allowed to grow on theses arms.

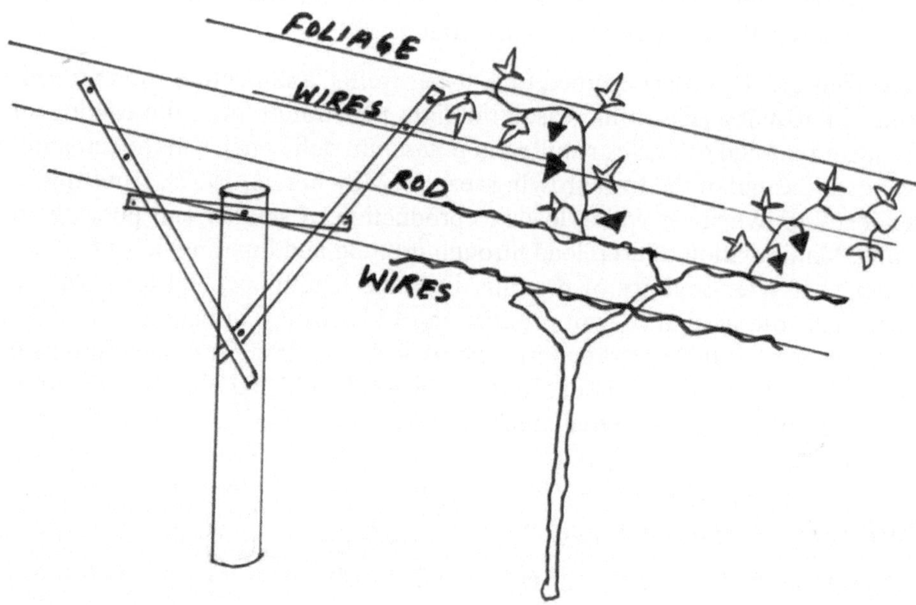

Fig. 40: Kniffen architecture

Pruning for canopy management

Pruning is most important operation in cultivation of Kiwifruit which should be done with great precision and care. Pruning and positioning of shoot improved light penetration and interception. The color and yield increase with precise pruning of kiwifruit. The vines grow 3-5m each year which become shabby, overcrowded and unmanageable, if not controlled by winter and summer pruning. Pruning is carried out in such a way that the fruiting areas are available every year requiring the wood to be young, this is achieved by following 3-4 years lateral replacement system which becomes a pruning cycle. In summer, fruiting shoots are headed back beyond 6-8 buds from the last fruit, if it has more growth again to be removed. Since male vine do not bear crops hence, all energy is diverted towards vegetative growth making plants vigours and huge stature which necessitates pruning regularly. A lateral is shortened by 5-7 buds in dormant condition and again the new growth to the same number of buds in the summer season. Overcrowding of lateral and arms should be avoided by removal of vertical shoots.

Partial defoliation reduced white rot incidence by 70% at the end of the summer. Leaf pruning has more effect on root growth than the fruit yield. Hence, root growth is more sensitive than fruit growth to reducing the source (leaf) to sink (fruit) ratio within the canopy. Major influences of canopy management on root growth indicated the importance of managing the root system for optimum performance of vines (Buwalda and Perrson, 1991).

Kiwifruit cv. 'Hayward' trained to a 'T-bar trellis' architecture, was pruned 5 times in growing season increased the light penetration in to the canopy, but frequent removal of leaves resulted in potassium deficiencies in the remaining leaves at the end of the fruit growing season (Miller *et al.*, 1997). Kiwifruit has tendency to over bear which leads to production of smaller and poor quality fruits. Manipulation of crop load through pruning and thinning helps to obtain better fruit size. Severity of pruning, i.e. cane length and bud load plays an important role in quality fruit production of kiwifruit (Galliano *et al*., 1999). Chandel *et al.* (2004) recorded highest yield of 'A' grade fruits and produced better size and quality without affecting the total yield and obtained maximum gross return from pruning treatment having a bud load of 75 canes with 6 buds per vine.

Leaf to fruit ratio management

Appropriats leaf to fruit ratio is essential to obtain export quality fruits in kiwifruit. Cooper *et al.* (1992) showed that 3:1 leaf to fruit ratio was necessary to support 40 fruits/m^2 canopy and greatest fruit size with export quality is obtained in cultivar 'Hayward'. Leaf to fruit ratio had a greater effect than crop load on fruit size and ratio of 3:1 or 2:1 is necessary for adequate return bloom in kiwifruit. Leaf to fruit ratio had great influence on the size of fruit, Lai *et al.* (1989) found that size of the kiwifruit on non-girdled laterals was not reduced at low leaf-fruit ratio (less than 2:1) because of translocation from other parts of the vine, could meet any shortfall in the supply of carbohydrate. Whereas, on girdled shoot minimum leaf-fruit ratio is required to support fruit growth in kiwifruit laterals without import of carbohydrate is 2:1 or approximately 225 cm^2 of leaf area per fruit. However, Snelger and Thorp (1988) observed that on girdled laterals final fruit weight increased with leaf area in a curvilinear fashion. Leaf area of 230 cm^2 and 335 cm^2 leaf area produced average sized fruit of 110 and 104 g respectively. On whole vines final fruit weights increased linearly with leaf area at rate of 5-6 g per 100 cm^2 leaf area over the range of 300-700 cm^2 per fruit in Hayward cultivars.

15

Canopy Development and Management of Ber

Ber (*Ziziphus mauritiana* L) ancient fruit of India grown in arid regions, 'king of arid zone fruits', thrive well in marginal areas. During severe hot, it shed off its leaves completely and thus conserves moisture by getting rid of transpiration mechanism. Flowering and fruiting mainly take place on secondary and tertiary branches. Ber require regular pruning to induce maximum number of healthy primary and secondary branches. It consists of hedging back of previous season's growth to ½ to ¾ and thinning out of weak, diseased and interwoven branches. This results good air circulation and light penetration inside the canopy. Ber flowers borne on the current shoot or new flush hence, it require annual and regular pruning. The best planting season is July-August under North Indian condition. However, plant without earth ball (bare rooted) can be planted during January-February.

Tree architecture development

Due to its bearing behavior, ber performed well in many tree architecture forms viz Open central leader, modified central leader, espalier trellis and bower architectures. According to Bal and Gill (2016), ber plants are best to plant at 7.5 m apart in square system in rain fed condition. When 'Umran' variety grafted on *Zizyphus mauritiana* rootstock, planted at 6 x 6 m, the composite tree remains semi-vigours in nature.

Canopy management in ber

Most of the ber cultivars are spreading in nature, hence soon after planting need to be supported during initial years for making proper shape. Bal and Gill (2016) reported that heading back of one fourth of the previous year's growth is necessary for heavy yield of quality fruits. 'Umran' variety of ber, pruned on 9th May by 6 buds resulted superior quality and 8 buds pruning in 3rd week of April at dormancy in 'Sanaur' variety proved best. In 'Umran' variety a selective pruning on 21st October to 5 November helped in improving the fruit quality, weight and palatability of fruits.

Reviving of old and unproductive orchard

Rejuvenation is an horticultural technique to make unproductive and senile tree productive by severe and corrective pruning. It is also used to change the top of the tree by top budding and grafting with superior variety. Bal and Gill (2010) reported that 25-30 years old ber trees required to rejuvenate for proper vegetative growth and fruiting. In rejuvenation trees are head back at main limb leaving 30 cm during mid May. Singh *et al.* (2015) carried out rejuvenation in old trees of 'Gola' and 'Umaran' variety planted at 6 x 12m spacing. In 'Gola', 3.55 times higher yield (36.4 kg/tree) with 78% of 'A' grade fruits were recorded. After sever pruning, 12-18 newly emerged shoots during mid August, growing out- wards, were required to keep as per tree vigour, and second thinning out to be carried out in September before beginning of flowering.

Canopy manipulation through pruning

Pruning is an important horticultural tools for canopy manipulation as well as for containment of plant spread, and sustainable growth and yield. Pruning intensities plays vital role for new shoot induction for maximum possible bearing surface production. Kumar and Ram (2009) studied pruning intensities in ber in horti-pastoral system and observed that medium pruning intensity, produced significantly higher yield, whereas, severely pruned trees produced less branches with vigours types. Severely pruned ber trees produced adventitious shoots in absence of pre-developed buds, hence, shoot sprouting was delayed.

Singh and Bal (2008) studied pruning in ber and reported that low pruning intensity improved the fruit yield and quality. The time of pruning determines the vegetative growth and canopy growth. Pruning in ber can be carried out from January to April to regulate fruit maturity for fruit availability time. In Maharashtra as well as western India, best pruning time was before end of April. Sandhu *et al.* (1992) and Kundu *et al.* (1992) noted the highest yield in 'Umran' variety pruned on 30th May. Dhaliwal and Sandhu (1982) reported that flowering intensity were higher in un-pruned trees while, fruit set and fruit retention were higher in pruned trees at 70 cm. Kundu *et al* (1995) recorded maximum fruit weight, where 50% branches were pruned from Origin point and maximum yield recorded with medium pruning intensity leaving 75 cm of the secondary shoots from base.

Dinesh (2002) studied the effect of pruning intensity in ber under semiarid condition and found highest fruit yield (28.94 kg per tree) in 50 per cent pruning intensity in 'Ponda Safeda' cultivars. Sharma *et al.* (2004) reported that eight years old ber trees have optimal growth, spaced at 25 x 25 feet distance, and light pruning was the best from yield (97.8 kg/tree) point of view. Anbu *et al.* (2009) standardized the planting density of ber and recorded higher yield per unit area at 8 x 3 m spacing.

Ber respond well to different pruning intensity, hence are suitable for high density planting system, practiced for increasing yield per unit area.

16

Canopy Management of Papaya

Papaya (*Carica papaya*) is a important delicious fruits of herbaceous in nature,. It is most precocious fruit crops cultivated as pure as well as intercropped with mango, guava, peach, and coconut. Papaya starts bearing in 8-10 months after planting. The acerage has increased tremendously recorded 6.2% growth rate annually. India is largest papaya producer, which contributes 42% of total world papaya production from 30% of the global area. Papaya is being cultivated over an area of 34,000 ha with 5940,000 MT production during 2017-18 (Anonymous, 2018). Uttar Pradesh, Bihar, Assam, Karnataka, Gujarat, Tamil Nadu and Andhra Pradesh, Maharashtra and West Bengal are leading in area and production of Papaya.

Papaya is one of the most appealing fruit due to precocious bearer and high money earning perennial herbaceous plant. There is vast scope of papaya in high density planting system. The pre-requisite of HDP is to develop varieties with short and erect petioles through breeding can probably assist in accommodation of more number of plants per unit area.

Some dwarf cultivars viz. 'Pusa Nanha', 'Ranchi Dwarf', 'CO1, 'CO-2, and 'Honey Dew-1' are being used for accommodating large number of plants per unit area.

High density planting system

Papaya is usually planted as filler crops with mango, guava and oranges. It is planted 2-3 m apart. However, 1.8×1.8m is normally followed for most of the varieties. 'Coorg Honey Dew' under South India conditions can be planted at 1.33×1.33 m which comprise 5609 plants per hectare. Dwarf papaya cultivars viz. 'Pusa Delicious' planted at 1.4×1.4 m or 1.4×1.6 m under subtropical condition gave higher yield.

Suitable varieties for HDP

1. Pusa Dwarf- It is dioecious, dwarf, statured, fruit oval n shape, precocious in nature.
2. Pusa Nanha- It is exceptionally dwarf, suitable for roof top culture and fit for HDP.

3. CO-1- Selected from 'Ranchi' seedlng population, dwarf, diocious, fruit round, to oval, orange, color flesh.
4. CO-6- Selection from 'Pusa Majesty', dioecious dwarf stature, fruit large in size.
5. Taiwan- It is gynodioecious variety with blood red flesh color and good taste.
6. Sunrise Solo: It is gynodioecious variety from Hawaii , pink fleshed color and good in taste.

Intercropping of papaya

Papaya being short statured plants suitable for cultivation with other perennial crops as intercrop. Papaya is a ideal intercrop in juvenile coconut due to early and continuous fruiting of papaya In South India (Nihad, 2019). Papaya cultivars of dwarf statured with short petiole leaves are suitable for high density planting system.

17

Canopy Development and Management of Bael

Bael (Angle marmelos) is an important arid fruits belongs to family Rutaceae and consists 2-3 species of which only one is in cultivation. It is important indigenous fruits of India, drought hardy tree, flowering and fruiting coincides with the onset of monsoon and fruits mature before commencement of summer. The tree shed off entire leaves during hot summer and fruits are exposed to direct sunlight, which favors ripening of fruits.

Bael is rich source of vitamin B2 (riboflavin) and minerals. This fruit also contain fair amount of antioxidant. Most of the Bael plantations are scattered with no systematic orcharding system. With increase in knowledge of its medicinal properties especially effective in dysentery ailments makes it a special fruits. Many Research Institutes have initiated research on germplasm conservation, varietal development maintenance and evaluation as well as production technology. ICAR-Institutes Central Institute of Arid Horticulture, Bikaner, Central Horticulture Experiment Station, Godhra, Gujarat, Central Institute for Subtropical Horticulture, Lucknow, Uttar Pradesh, have initiated work on cultivar development, production technology (high density planting system, canopy management practices, the results obtained are quite encouraging.

Bael cultivars suitable for tree architecture development

CISH Bael 1: Plants are medium in height with low spreading type fruits are oval oblong having size 15-17 x 39-41 cm, total soluble solids 35-40 'degree brix' fruit weight 800-1200 g/fruit, thin rind (0.10-0.12 cm). Average yield per tree is 50-60 kg and suitable for close planting system.

CISH Bael 2: The plants are semi vigorus and more spreading type, fruits are round or oblong shape, 1800-2700 g/fruit, with 37-410 Brix total soluble solids. Fruits are good in taste, yield 40-50 kg/tree.

Pant Aparna: Plants are dwarf with sparse foliage, drooping branches, less thorny, precocious and high yielding.

Pant Shivani- Plants are upright dense, growth, good keeping quality.

Tree architecture

Usually bael seedling plantations are large in size 20-30 feet height, 4-5 meter spread. With the standardization of successful propagation technique lead the plants with small stature. The budded plants starts early flowering and fruiting and thus trees are small in height and spread as compared to seedling trees.

The bael grafts are planted in July-August after proper establishment in main field, the main stem is cut back leaving 90-100 cm this will encourage number of new shoots, allow 3-4 well spaced out growing bud and remove remaining new shoots. Thus 3-4 primary scaffolds branches will develop and matured in 8-10 months. In next growing season prune the primary scaffolds by 50 percent of the length on each primary branch, 3-4 secondary branches will grow and remove the remaining ones. The complete canopy will develop in 4-5 years, after that flowering and fruiting starts.

Pruning

Bael do not require regular pruning, the flowering in bael starts in July-August and fruits mature in April-May. Only criss-crossed, overcrowded, diseased, dried, broken branches/twigs should be removed. If bearing tree has become over-crowded and central canopies are not receiving proper light interception, center opening should carried out, apply Chaubattia paste or copper oxy-chloride paste on the cut portion immediately

References

Anonymous (2018). "Horticulture At a Glance" Area, production and productivity of horticultural crops. Department of Horticulture and Food Processing, Govt of Uttar Pradesh, Lucknow.

Anbu S., Balasubramanyan, S., Venkatesan, K; Selvarajan M, and Duari singh R. (2009). Evaluation of varieties and standardization of production technology in Ber (Zizyphus mauritiana) under rainfed vertisols. Acta Horticulturae. 840: 55-60. Annual IDFTA. Conference, Nigara Falls, New York, USA.

Anonymous (2019) Canopy reorientation in overcrowded in productive mango under hdpc crop production. In ICAR-CISH annual report 2019. Published by Director, ICAR-CISH, Lucknow.

Acock B. and Allen L.H. Jr. (1985). Crop response to elevated carbon dioxide concentration. Indirect effects of increasing carbon dioxide on vegetation DOE/ER-0238. B.R. strain and J.D. Cure (eds). US Department of Energy. Carbon Dioxide. Res. Div. Washington D.C. pp. 53-97.

Archer, E. (1987). The role of light and canopy management in south African Vineyards. Deciduous Fruit Grower, 37(10): 397-405.

Assaf. R. (1988). Training methods and planting densities for pear trees in hot countries. Fruits Paris, 43 (2): 113-125.

Avery, D.J.(1957). Effect of climatic factors on the photosynthetic efficiency of apple leaves. In: Climate and Orchard, edited by Pereira, H.C. comm. Agr. Res. Rept. No. 5 pp. 25.31.

Awasthi, R.P. and Mehta, Kuldeep. (2000). Strategies for developing high density plantings in Horticultural crops, In: Souvenir of National Seminar on Hi-tech Horticulture, Bangalore, 26-28. June, pp. 29-33.

Bal J.S. and Gill. G.S. (2016). Canopy management, nodal pruning and planting density in Indian Jujube (ber). Acta Horticulturae. 1116:99-104.

Baldwin, J.B.(1964). The relation between weather and fruitfulness of Sultana Vine. Aust. J. Agri. Resh., 15: 920-928.

Bannett. J. and Turner, R. (1978). Rept. Long Ashton Res. Sta for 1977, pp.29.

Barden, J.A. (1977). Apple tree growth, net photosynthesis, dark respiration, and specific leaf weight affected by continuous and intermittent shade. J.Amer.Soc.Hort.Sci., 102: 391-394.

Barden, J.A. (1978). Apple leaves, their morphology and photosynthetic potential. Hort. Science, 13(6): 644-646.

Barritt, B.H., Rom, C.R., Konishi, B.J. and Dilley, M.A. (1991). Light level influences spur quality and canopy development and light interception influence fruit production in apple. Hort. Science, 26(8):993-99.

Bevington K.B., Falivene S., Moulds G., and Krajewski, A.J. (2002). Pruning citrus for export fruit size. CMDG Final report Project 50. New South Wales Dept of Agric Australia.

Bist, L.D., Rai, N. and Sharma, N.L. (1992). Effect of plant growth regulations on tree shaping and productivity of low chilling pear cultivar Gola. In: Proceeding of Golden Jubilee National Seminar on Emerging Trends in Temperate Fruit Production in India, at D.Y.S. Parmar University of Horticulture and Forestry, Solan. pp.39.

Blake, A.A. (1917). Observations upon summer pruning of the apple and peach. Proc. Am. Soc. Hort.Sci., 14:14-23.

Blake, M.A. (1916). The second season with the peach orchard. New jersey Agric. Exp. Sta. Bull., 231:28-31.

Blinovski, I.K. (1970). The effect of shoot angle on photosynthesis in apple. Dokl. Mark. Sel. Joz. Acad.K.A. Timirjazeva., 158: 135-41.

Buban, T. and Faust, M. (1982). Flower bud induction in apple tree: Internal control and differentiation. Hort. Rev., 4:174-203.

Buban, T. and Faust, M. (1982). In: Horticultural Reviews (ed. J.Janick) Val.4, AVI Pub. Co. Westport.

Bukovac, M.J. (1985). Spray application technology: shortcoming and opportunities with special reference to the tree fruits. Proc.Conf. Agr. Resh. Inst. Bethesda. Md., p. 25-38.

Burondkar M.M., Gunjate R.T., Magdum M.B. and Govekar M.A. (2000). Rejuvenation of old and overcrowded Alphonso mango orchard with pruning and use of Paclobutrazol . Acta Horticulture. 509: 681-686.

Burondkar, M.M. and Gunjate, R.T. (1991). Regulation of shoot growth and flowering in Alphanso Mango with poclobutrazol. Acta Hort., 291: 79-84.

Burondkar, M.M. Gunjate. R.T., Magdum, M.B., Govekar, M.A. (2001). Rejuvenation of old and over crowded Alphonso mango orchard with pruning and use of paclobutrazol. Acta Hort. 509.

Buttrose, M.S. (1969). Fruitfulness in grape vines: Effect of light intensity and temperature. Bot. Gaz., 130:166-173.

Buwalda, J.G. and Persson, H. (1988). Root growth of kiwifruit vines and the impact of canopy manipulations. Proceeding of an ISSR Symposium. 431-441.

Campbell. R.J. and Wasieleweki (1999). Mango tree training techniques for the hot tropics. Acta Hort., 509: 641-651.

Cannell M.G.R. and Kimeu B.S. (1995). Uptake and distribution of macronutrients in trees of Caffea arabica in Kenya as affected by sectional climatic differences and the presence of fruits. Ann. Applied Biol. 68: 213-230.

Cantin C.M., Pinochet J., Gogorcena Y., Moreno M.A. (2010). Growth yield and fruit quality of Van and Stark Hardy Giant were cherry cultivars as influenced by trafting on different rootstocks. Scientia Horticulturae, 123: 329-335.

Chacko, E.K. (1968). Studies on the physiology of the flowering and fruit growth in mango (Mangifera indica L.) Ph.D. thesis, IARI, New Delhi.

Chacko, E.K. and Ananthanarayan, T.V. (1982). Accumulation of reserve substances in Mangifera indica (L) during flower initiation. Z. Pffanzen-physiol. Bd., 1065:281-285.

Chalmers, D.J., and B. van den Ende. 1975. J. Agr. Victoria 73:479-476.

Chandel, J.S., Bharti, O.A. and Rana, R.K. (2004). Effect of pruning severity on grwth, yield and fruit quality of Kiwifruit (*Actinidia deliciosa* Chev.). Indian J. Hort., 61(2): 114-117.

Charbonneau, A. and Lespinasse, J.M. (1989). Influence of canopy management on micro-climate and photosynthesis first consequences on apple production. Acta Hort., 243:185-194.

Chartzoulakis K.I., Therios and B. Noitsakis. (1993). Effect of shadinhg on gas exchange, specific leaf weight, and chlorophyll content in four kiwifruit cultivars under field conditions. Journal of Hort Science. 68:605-611.

Cline, M. G. (2000). The role of hormones and apical dominance. New approaches to an oldproblem in plant development. Physiol. Plantar. 90:230–237.

Child, R.D. and Atkins. H. (1978). Crops from the 1972/73, meadow orchard. Rpt. Long Ashton Res. Sta. for 1977; p. 28-29.

Chittirai Chelvan, R., Shikhamany. S.D. and Chadha, K.L. (1985). Constitution of the leaf area towards the bunch development in Thompson Seedless grapes. Indian J.Hort., 42: 156-150.

Chootummatat, V., Turner, D., & Cripps, J. E. L. (1990). The water use of plum trees (Prunus salicina) trained to four canopy arrangements. Scientia Horticulturae, 43, 255-271.

Christopher, E.P. (1976). In: Pruning Manual. Macmillan Co.Inc. NY, p.320.

Cmelik z., Duralija B., Bencic D., Druzic J. (2002). Influence of rootstocks and planting density on performance of plum tree. Acta Horticulturae. 557: 307-310.

Cooper, K.M. and Marshall, R.R. (1992). Crop loading and canopy management. Acta Hort., 297: 501-508.

Costes, E., Lauri, P.E., and J.L. Regnard (2006). Analysis of fruit tree architecture: Implications for tree management and fruit production in Horticultural Reviews (Ed. Jules Janick), Published by John Wiley & Sons, Vol 32 pp. 1-32.

Dann, I.R. and Jerie, P.H. (1988). Gradients in maturity and sugar levels within peach trees. J. Amer. Soc. Hort. Sci., 113: 27-31.

Dann, I.R., Mitchell, P.D. and Jerie, P.H. (1989). The influence of branch angle on gradients of growth and cropping within peach trees. Sci. Hort. 43:37-45.

Das, Vikash and Rana, B.R. (2013). Effect of Canopy management on growth and yield of mango cv. Amrapali planted at close spacing. Journal of Food Agriculture & Environment. 11(1): 316-319.

Dash Bikash, Mehta Sarita, Singh S.K., Mali Santosh, Dhakar M.K. and Singh A.K. (2018). Pruning effects on Sardar guava planted in ultra high density orcharding under different rainfall scenario. Indian Journal of Horticulture 75(4): 583-590.

Davenport, T.L. and .Nunez- Elisea, R. (1997) Reproductive physiology pp 69-146; In : R.E. Litz (ed). The mango: Botany, production and uses. CAB International, walling ford U.K.

Day, K.R., Dejong, T.M., and Hewitt, A.A. (1989). Post harvest and preharvest summer pruning of "Fire Brite' Nectarine Trees. Hort. Science, 24(2): 238-240.

De Salvador, F.R. and Dajong. T.M. (1989). Observation fo sunlight interception and penetration into the canopies of peach trees in different planting densities and pruning configuration. Acta.Hort., 254: 341-346.

Debnath S., Bauri, F.K., Bandyo –padhyay B., Misra D.K., Mavdal, K.K., Murmu I. , and Patil P.C. (2015). Identification of optimum Leaf area index for high density planting of Banana cv. Martman in gangetic alluvium region of West Bengal. Journal of Crop and Weed. 11 (2): 63-66.

Debnath S., Barui, F.K., Bandopadhyay, B., Mishra, D.K. Mandal, K.K., Murmu and Patil, P. (2015). Identification of optimum Leaf area index (LAI) for high density planting of banana cv. Martman in gangetic alluvium region of West Bengal. Journal of crop and weed. 11(2): 63-66.

DeJong T.M., Day K.R., Doyle J.F. and Johnson, R.S. (1994). The Kemey Agricultural Center perpendicular "V" orchard system for peachesv and nectarins. Hort Technology 4:362-367.

Duncan, W.G., Shaver, D.N. and Willaiams, W.|A. (1973). Insolation and temperature effects on maize growth yields. Crop Sci., 13:187-90.

Durand, G. (1997). Effects of light availability on the architecture of canopy in mango cv. Manzana trees. Acta Hort., 455: 217-227.

Dzhincharadze, G.D. and Dzahabnidze, G.K. (1989). The effect of different pruning methods on the productivity and biochemical properties of mandarin fruit. Sub. Trop. Kul't No. 2:83-86.

Edward, J.C. and Gaurishankar (1964). Rootstock trial for guava (*Psidium guajava* L.) . Allahabad Farmers. 38: 249-250.

Elfving, D.C. and McKibbon, E.D. (1992). Tree support another management tool of apple growers. Compact Fruit Tree, 35th

Faust M and Timon B. (1995). Origen and dissemination of peach. Horticultural Review 17: 331-379.

Fejes, S., E_Horn, and T. Brunner. 1992. Gyumolcssoveny (Fruit trellis). Budapest: Mezogazdasagi Kiado.

Ferree, D.C. (1980). Canopy development and yield efficiency of 'Golden Delicous' apple tree in four orchard management systems. J. Amer. Soc. Hort. Sci., 114:869-875.

Ferree, D.C. (1983). Managing light in high density orchards. Proc. Wash.State Hort. Assn. 79:129-133

Ferreira de souse, C.A., Gondim cavalcanti Maria Irisvalda leal, vasconcelos Lopes . Lucio Flavo, Sousa Humberto umbeilo, Ribeiro valdenir Queiroz (2012). Tommy Atkins mango trees subjected to high density planting in sub humid Tropical climate in northern Brazil. Pesq. Agropec, Bras, Brasilia. 47(1): 36-43.

Fivaz, J. and Stassen, P.J.C. (1996). Year South Afr. Mango Growers, Association, 16: 32-35.

Flore, J.A. (1994). Stone fruits. In: Handbook of Environment Physiology of Fruit Crops, Vol. I: Temperate crops, pp 230-270. Schaffer, B. and Anderson, P.C. (Eds), CRC Press. Boca Raton.

Flore, J.A. and Lakso A.N. 1989. Environmental and physiological regulation of photosynthesis in fruit crops. Hort Rev. 11:111-157.

Forshey, C.G. and Mc Kee, M.W. (1970). Production efficiency of a large and a small Mc Intosh apple tree. Hort Science, 5: 164-165.

Fotzgerald, J. and Patterson, W.K. (1994). J. American Soc. Hort. Sci., 119 (5): 893-898.

Gabrielson, E.K. (1984). The influence of light of different wavelengths on photosynthesis in foliage leaves. Physiol. Plant, 1: 113-123.

Gallagher, J.N. and Biscoe, P.V. (1978). Radiation absorption, growth and yield of cereals. J. Agr. Sci. Camb., 91: 47-60.

Galliano, A. Youssef, J., Tonutti, P. and Giuliva, C. (1999). Effect of summer pruning of kiwifruit on yield. Acta Hort., 282: 127-132.

Gavrilescu E., Cosmulescu S., Baciu A., Botu M. (2004). Influence of cultivar- rootstock combination on dinamical physiological process in plum and genetics, Breeding and Pomology . Book of Abstracts 8(112E).

Gerdts, M. (1987). Training and pruning for maximum production and quality. Proc. National Peach Council Conv., 42:122-124.

Gogoi, B., Khangia. B. and Baruah K.(2015). Effect of high density planting and nutritient management on growth and yield of Banana cv. Jahaji (Musa AAA). Progressive Horticulture. 47 (2): 208-212.

Goswami, A.M., Saxena, S.k and Kurein, S. (1993). High density planting in citrus. In: Advances in Horticulture. Edited by. Chandha, K.L and Pareek, O.P., Malhotra publishing House, New Delhi pp. 645-658.

Groza, T. (1972). Results obtained by modifying the crown form of apple tree. Revista de Horticultura si Viticultura, 21 (10): 33-38.

Grzyb Z.S., Rozpara E., (1998). Plum production in Poland. Acta Horticulture, 478: 19-22.

Gubler, W.D., Marois, J.J., Bledsoe, A.M. and Bettiga, L.J.(1987). Control of Botrytis bunch rot of grape with canopy management. Plant Disease, 71(7): 599-601.

Gupta, N.K. and Bist, L.D.(2005). Effect of different planting systems and paclobutrazol on vegetative growth of Bagugosa pear. Indian J. Hort., 62(1): 20-23.

Hall, D.O., Scurloc, J.M.O, Bolhar-Nordenkamp, H.R., Leegood R.C., and lo, eds. (1993). Photosynthesis and production in a changing environmental (London, UK, Chapman and Hall).

Hampson Cheryl R., Qfuamme Harvey A, Kappel Frank and Brownlee Robert T. (2004). Varying density with constant Rectangularity : Effect on apple tree growth and light inter ception in three training system. Hort Science. 39 (3): 501-506.

Hamzakheyal, H., Ferree, D.C. and Hartmann, F.O. (1976). Effect of lateral shoot orientation on growth and flowering of young apple trees. Hort. Sci., 11:393-396.

Hansen. P. (1969). ^{14}C studies in apple trees. IV. Photosynthetic consumption in fruits in relation to leaf-fruit ratio and the leaf fruit position. Physiol. 22: 186-198.

Heinicke, D.R. (1963). The micro-climate of fruit trees. III. Foliage and light distribution patterns in apple trees. Proc. Amer. Soc. Hort. Sci., 83:1-11.

Heinicke, D.R. (1964). The microclimate of fruit trees. III. The effect of tree size on light penetration and leaf area in Red Delicious appe tree. Prc. Amer. Soc. Hort. Sci., 85:33-41.

Heinicke, D.R. (1966). Characteristics of Mc Intosh and Red Delicious apple as influenced by exposure to sunlight during the growing season. Proc. Amer. Soc. Hort. Sci., 89:10-13.

Heinicke, D.R. (1975). High density apple orchards planning, training and pruning. U S. dept. Agr. Hdbk.458.

Holland, D.A.(1968). The estimation of total leaf area on a tree. Rept. East Malling Res. Sta. for 1976, pp.101-104.

Hrotko K., Magyar L., Simon G., Klenyan T., (1998). Effect of rootstocks on growth of plum cultivars in a young orchard. Acta Horticulturae 478: 95-98.

Hudson, J.P. (1970). Maximum intensity orchards, growing fruit without trees. Proc. Amer. Soc. Hort. Sci., 89:10-13.

Hudson, J.P. (1970). Maximum intensity orchards, growing fruit without tree. Proc.18th Intl. Hort. Congress, 1:56.

Huett, D.O. (2004). Macadamia physiology review: a canopy light response study and literature review. Australian J. Agri. Resh., 55(60): 609-624.

Hunter, J.J. (1998). Plant spacing implications for grafted grape vine. II soil, water, plant, water relations, canopy physiology, vegetative and reproductive characteristics, grape composition, wine quality and labour requirements. South African J. Enology Viticulture, 19(2):81-91.

Isarva, I.S., Pereyaslova and I.D. Zelepukhin, (1983). Role of rootstock in apple yield formation. Nauchnye Doklady Vysshei Shkoly Biologi Cheskie Nauki, 7: 97-100.

Ito, By. A., Yoshioka, H., Hayama, H. and Kashimura, Y. (2004). Reorientation of shoots to the horizontal position influences the sugar metabolism of laterals buds and shoot internodes in Japanese pear (*Pyrus pyrifolia* Burm. Nak.). J. Hort. Sci. Biotech., 79(3): 416-422.

Jackson J.E. (1989). World-wide High density planting in resca. Acta Hort. 243:17-27

Jackson, J.E. (1978). Utilization of light resources by high density Planting systems. Acta Hort., 65:61-70.

Jackson, J.E. (1980). Light interception and utilization by orchards systems. Hort. Rev., 2:208-287.

Jackson, J.E. (1982). Light interception and utilizatrion by orchard systems. Horticultural Reviews. Vol. 2 edited by Jules Janick, AVI, Publishing Company, INC. Westport, Connecticut, pp.209-257.

Jackson, J.E. and Palmer, J.W. (1972). Interception of light by model hedgerow orchards in relation to latitude, time of the year and hedgerow configuration and orientation. J. Applied Ecol., 9: 341-358.

Jackson, J.E. and Palmer, J.w. (1980). A computer model study of light interception by orchards in relation to mechanized harvesting and management. Scientia Hort., 13: 1-7.

Jackson, J.E. Palmer, J.W., Perring. M.A. and Sharples, R.O. (1977). Effects of shade on the growth and cropping of apple trees. III effects on growth chemical composition and quality at harvest and after storage. J.Hort. Sci., 52:267-282.

Jackson. J.E. and Beakbane, A.B.(1970). Structure of leaves growing at different light intensities within mature apple tree. Rept. East Malling Res. Sta., 1969:87-89.

Kalra, S.K., Sidhiv, P.I.S., Dhaliwal, J.S. and Singh, R. (1994). Effect of different spacing on yield and quality of guava cv. Allahabad Safeda. Indian J. Hort., 51(3):272-274.

Kappel, F. and Brownlee, R. (2001). Early performance of Conference pear on four training system. Hort. Sci., 36(1):69-71.

Kata, T. and Ito, H. (1962). Physiological factor associated with the shoot growth of apple trees. Tohoku J.Agric. Res., 13:1-21.

Kim, J.K., Kim, S.b., Kim, K.Y., Cho, M.D and Kim J.H. (1987). The effect of planting density on tree growth, yield and root distribution in a pear orchard. Res. Rep. Rural Der. Adm. Hort. Korea Republic, 29(1): 93-98.

Kliewer, M. and Bled Soe, A. (1987). Influence of hedging and leaf removal on canopy micro climate, grape composition and wine quality under California conditions. Acta Hort., 206:157-168.

Koblet, W., Keller, M. and Candolfi-Vasconealos, M.c. (1997). Effect of training system, canopy management practices, crop load and rootstock on grapevine photosynthesis. Acta Hort., 427: 133-140.

Koroid, A.S. (1987). Sadovodstvo, 3:16-17.

Krajewski A.J. (2010). Pruning of citrus in Southern Africa: a Hacker's guide. Citrus J. 6(4): 19-23.

Krajewski A.J. and Pittaway T., (2000). Manipulation of citrus flowering an fruiting by pruning. Proc. Intl. Soc. Citri. 357-360.

Kriedmann, P.E. and Smart, R.E. (1971). Effect of the irradiance temperature and leaf weather potential on the photosynthesis of vine leaves. Photosynthetica., 5:6-15.

Kudryavets, R.P. and Trusov, V.P. (1975). The effect of different pruning methods on the light regime and photosynthesis in densely planted apple trees. Sbornik Nauch Rabot N.I. Zonal'n Int.

Kumar Sunil and Ram S.N. (2009). Performance of ber (*Ziziphus mauritiana* L.) based horti-pastoral system as influenced by pruning intensities, pastures and weather condition. Indian Journal of Agroforestry 11 (2): 14-19

Kumar, Dinesh (2002). Effect of pruning intensity on vegetative growth and yield of Indian Jujube (*Z. mauritiania*) under semi-arid condition. Indian Journal of Agricultural Sciences. 72 (11): 659-660.

Kumar, J., Thakur, G.C., Verma, H.S. and Rehalia, A.S. (1992). Effect of different pruning intensities on growth yield and fruit quality of plum cultivars Santa Rosa. In: Proceeding of Golden Jubilee National Seminar on Emerging Trends in Temperate Fruit Production in India, at D.Y.S. Parmar University of Horticulture and Forestry, Solan. pp.40.

Kundi , S.S. (1995). Haryana J. Hort. Sci. 24 (1):

Kurian, R.M., Reddy, V.V.P. and Reddy, Y.T.N. (1996). Growth, yield, fruit quality and leaf nutrient status of thirteen year old, 'Alphonso' mango trees on eight rootstocks. J. Hort, Sci., 71: 181-186.

Ky I., Lorrian B., Jourdes M., Pasquier G., Fermaud, M., Geny L., Rey P., Doneche B. and Teissedre P.L. (2012). Assessment of gray mold (*Botrytis cinereal*) impact on phenolic and sensory quality of Bordeaux grapes musts and wines for two consecutive vintages. Aust J. Grape Wine Res. 18:215-226.

Lai, R., D.J. Woolley and G.S. Lawes (1989). Effect of leaf to fruit ratio on fruit growth of kiwifruit (*Actinidia deliciosa*). Scientia Horticulturae., 39: 247-255.

Lakso T.N. and Robinson T.I. Intensive orchard system management of demand and particionong in apple. Acta Hort 451:405-415.

Lakso, A.N. and Seeley, E.J. (1978). Environmentally induced response of apple tree photosynthesis. Hort. Science, 13: 646-650.

Lakso, A.N.(1975). Light studies in apple trees. New York Food and Life Sci. Quart., 8(4):6-8.

Lakso, A.N., Robinson, T.L. and Carpenter, S.G. (1989). The palmette leader: A tree design for improved light distribution. Hort. Science, 24: 271-275.

Lal B., Rajput, MS., Rajan S., and Rathore D.S. 2006. Effect of pruning on rejuvenation of old mango trees. Indian Journal of Horticulture 57:240-242.

Lal, B., Rajput, M.S. and Rathore, D.S. (2000). Effect of pruning on rejuvenation of old mango trees. Indian J. Hort., 57(3): 240-242.

Lanauskas J., Uselis N., Kvklys D., Buskiene L. 2012. Rootstock effect on the performance of sweet cherry cv Lapins. Horticultural science (Prague) 39: 55-60.

Larsen, F.E., Fritts, R., Olsen, Jr. and Olsen, K.L. (1986). Rootstock influence on delicious and Golden Delicious apple fruit quality at harvest and after storage. Scientia Horticulturae, 26: 339-49.

Larsen, F.E., Olsen, K.L. and Fritts, R. Jr. (1985). Rootstock effects on apple maturity studied. Good Fruit Grower, 37: 46-47.

Lauri P.E. and ClaverieJ. (2005). Sweet cherry training to improve fruit size and quality. An overview of some recent concepts and practical aspects. Acta Horticulturae 667: 361-366.

Lauri PE and Corelli Grappadelli, I. (2014). Tree architecture, flowering and fruiting -Thoughts on training pruning and ecophysiology. Acta. Hort. 1058, 291-298.

Lespinasse, J.M. (1989). A new fruit training system: The 'Solen'. Acta Hort., 243:117-120.

Lespinasse, J.M. and Delort, J.F. (1986). Apple tree management in vertical axis: appraisal after ten years of experiments. Acta Hort., 160: 139-155.

Loomis, R.S. and Gerakis, P.A. (1975). Productivity of agriculture ecosystems. In J.P. Cooper (ed.) photosynthesis and productivity in different environments. Cambridge Univ. Press, London. Pp. 145-172.

Looney, N.E. (1968). Light regimes in standard size apple trees as determined spectrophotometrically. Proc. Amer. Soc. Hort. Sci., 93:1-6.

Luckwill, L.C. (1968). The effect of certain growth regulators on growth and apical dominance on young apple trees. J. Hort. Sci., 43:91-101.

Machado, D.L.M., Alves, R.R., Siqueira, D.L., Salomao, L.C.C. and Silva, S.D.R. (2013). Development and production of Tahiti Lime IAC-5 grafted on 'Flying Dragon' (*Poncirus trifoliate* var. *Monstrasa*) grown in high planting densities. Paper presented in 12th Intn. Citrus Cong, Valencia, Spain, 18-23 Nov, pp181.

Madhavarao, V.N. and Shanmugavelu, K.G. (1975). Pruning- A new technique for mango growers. Agriculture Agro. Industries J., September. P.5-7.

Madhavarao, V.N. and Shanmugavelu, K.G. (1976). Studies on the effect of pruning in mango. Pog. Hort., 8:21-28.

Majumdar, P.K. and Sharma, D.K. (1985). 2nd International Symposium on Mango, India. Abst No 3.9, p. 20.

Mallick S.K., Mitra S.K., Banik B.C., Sen S.K. and Bose T.K. (1985). Induction of flowering in mango cv Fazli with chemical treatments. 2nd International Symposium on Mango, Bangalore, India pp 37.

Marini, R.P. and Barden, J.A (1982). Yield, fruit size and quality of three apple cultivars as influenced by summer and dormant pruning. J. Amer. Soc. Hort. Sci., 107(3): 474-479.

McFadyen, L.M., Morris, S.G., Oldham, M..A., Huett, D.O., Meyers, N.M., wood. J. and Mc Conchie, C.A. (2004). The interrelationship between orchard crowding, light interception and productivity in macadamia. Australian J. Agri. Resh., 55(10): 1029-1038.

Mckenzie, D.W. (1972). Intensive orchards in New Zealand. The Orchardist of N.Z., 44:175-181.

Mckenzie, D.W. and Rae, A.N.(1978). Economics of high density apple production in New Zealand. Acta Hort., 65:41-46.

Mika., A. and Antoszewski, R. (1972). The effect of leaf positioning and tree shape on the rate of photosynthesis in the apple tree. Photosynthetica, 6:381-386.

Miller, S., Broom, F., Thorp, G. and Barnett, A. (1997). Kiwifruit canopy management-transition to leader pruning. Orchardist, 70(8): 39-41.

Miller, S.S.(1982). Regrowth, flowering and fruit quality of 'Delicious' apple tree as influenced by summer pruning. J. Amer. Soc. Hort. Sci., 107:975-978.

Milosevic Tomo; Milosevic Nebojsa. Milivojevic, Jelena; Glisic Ivan and Nikolic Radmila (2014). Experience with Mazzard and colt sweet cherry rootstock in serbia which are used for high density planting system under heavy and acids soil conditions. Scientia Horticulturae 176: 261-272.

Nath, V. and Pongener, Alemwati (2017). Canopy architecture management for doubling India's fruit production , in: Doubling Farmers Income Through Horticulture (ed.) K.L. Chadha, S.K. Singh, P Kalia, W.S. Dhillon , T.K. Behra, Jai Prakash) pp; 191-204.

Mika, A. (1971). Proc. Inst. Sad. 15:63-72.

Mika A. (1975). Fruit Sci. Rept. 2(1): 31-42

Mika A. (1982). Proc. XXI Inter. Hort. Congress, Hambur, 29 Aug – 4 Sep, 1: 209-221, .

Mohammad, S., Wilson, L.A. and Predergast, N. (1984). Guava meadow orchard: effect of ultra-high density plantings and growth regulators on growth, flowering and fruiting. Tropical Agri., 61(4): 297-301.

Monteith J.L. (1977). Climate and efficiency of crop production in Britain. Trans . Phil R. Soc. London B. 281: 277-292.

Morgan, D.C., Stanley, C.J., Volz, R. and Warrington, I.T. (1984). Summer pruning of Gala apple: the relationship between pruning time, radiation penetration and fruit quality. J. Amer. Soc. Hort. Sci., 109:637-642.

Morris, J.R., Sims, C.A. and Cawthon, D.L. (1985). Yield and quality of 'Nigara' grapes as affected by pruning severity, nodes per bearing unit, training system, and shoot positioning. J. Amer. Soc. Hort. Sci., 110: 186-191.

Mullins, M.G. (1965). The gravitational responses of young apple trees. J. Hort. Sci., 40:237-47.

Mundy D.C., Agnew R.H., Wood P.N. 2012. Grape tendrils as an inoculum source of Botrytis cinereal in vineyards – a review .N Zealand Plant Protec. 65:218-227.

Myers, S.C. and Ferree, D.C. (1983). Influences of summer pruning and tree orientation on net photosynthesis, transpiration, shoot growth and dry weight distribution in young apple trees. J. Amer. Soc. Hort. Sci., 108:4-9.

Nalina I., Kumar N. and Sathiamoorthy, S. (2000). Studies on high density planting in banana cv Robusta (AAA) influence on vegetative characters. Indian J. Hort 57:190-95.

Nihad, K; Krishnakumar, V; Rojkumar M.; haris, A-A and Bhat R. (2019). Papaya inter cropping-an income for newly planted coconut gardens. India Coconut Journal. 62(4):27-29.

Nath V. and Pongener Alemwati (2017). Canopy management for doubling India's fruit production, in Doubling Farmers Income, Through Horticulture (ed: Chadha K.L., Singh S.K., Kalia P., Dhillon W.S., Behera T.K. and Singh J. P.) Daya publishing House. pp; 191-204.

Oosthuyse, S.A. (1993). Evaluation of winter pruning to synchronize the flowering of sensation mango trees, Year Book-South African Mango Growers, Association, 13:54-57.

Palmer, J. W. and Jackson, J. E. (1977). Seasonal light interception and canopy development in hedgerow and bed system apple orchards. J. Applied Ecol., 14:539-549.

Palmer, J.W. (1988). Annual dry matter production and partioning over the first 5 years of a bed system of crispin/M-27 apple trees at four spacing. J. Apple. Ecol., 25:569-578.

Parnia, C., Dumitrescu, G., Rosca, C. and Prahoveanu, G. (1986). Acta Hort., 160 pp.167-176.

Park W.J. (2011). Plant regulatory signal debates and Perspectives. Journal of Plant Biology. 54: 143-149.

Pathania N.S., Sehgal O.P. and Gupta Y.C.(2000). Pinching for flower regulation in Sim Carnation. Journal of Ornamental Horticulture. 3:114-117.

Peng, Fan Ren, Huang, B., Li, Jie, Zang, Jilin, Peng, F.R., Huang, B.L., Li, J. and Zhang, J.L. (1999). The pattern of light distribution and utilization on compound pear orchard in sea coast area. J. Plant Resource Env., 8(4): 25-29.

Pepelyankov, G. and Grnevski, V. (1986). Effect of planting density on the growth, fruiting and fruit quality of pear cultivar Williams Bon Chretien on Quince rootstock. Resteniev dni Nanki. 23(11):91-96.

Pereirade Bem Betina, Bogo Amauri Everhart Sydney, Casa Recardo Trezzi, Goncalves Mayra Juline, Fieho macron, Jose Luiz and Cunha da Isabel cristina (2015). Effect of 'y' Trellis and vertical shoot positioning training systems on downy mildew and Botytis bunch rot of grape in high lands of Southern Brazil. Scientia Horticulturae . 185: 162-166.

Perez Harvey, J., Fuenzalida, M., Cornejo, P. and Momberg, W.(1987). Influence of canopy management, crop load and ringing on the incidence of stalk necrosis and on berry quality in cultivar Sultania (Thompson Seedless). Ciencia-e-Investigacion-Agaria, 14(2): 97-106.

Pezet R., Viret O., Perrett C., Tabacchi, R. (2003). Latency of Botrytis cinrea Pers: Fr and biochemical studies during growth and ripening of two grape berry cultivars respectively susceptible and resistance to gray mould. J. Phytopathology 151: 208-214.

Porpiglia, P.J. and Barden J.A. (1980). Seasonal trends in net photosynthetic potential, dark **respiration and specific leaf weight of apple leaves as affected by canopy position. J. Amer.** Soc. Hort. Sci., 105: 920-923.

Prasad, A. 1966. A rootstock trial in guava (*Psidium guajava* L.) proceedings, All India Symp. Hort., cited by Sinha et al. (1993). In: Advance in Hort., Vol.2.pp551-562 edited by Chadha, K.L. and O.P. Pareek, Malhatra Publishing House, New Delhi.

Rajbhar, Y.P, Singh, S.D., Lal Mohan, Singh Gopal and Rawat P.L. (2016). Performance of high density planting of mango (*Mangifera* India) under mid-western plain zone of Uttar Pradesh. Int. J. Agric Sci. 12 (2): 298-301.

Ram, S. (1990). Hormonal physiology of flowering in Dashehari Mango. J. App. Hort., 1(2): 84-88.

Ram, S. (1996). High density orcharding in mango. Research Bulletin No. 122, Directorate of Experiment Station, G.B.P. U.A.T; Pantnagar, (UA).

Ram, S. and Sirohi, S.C. (1985). 2nd International Symp. on mango, India, Abst. No. 4. 21, p.38.

Ram, S. and Sirohi, S.C. (1989). Feasibility in high density orcharding in Dashehari mango. Acta Hort., 1: 14-17.

Ram, S. and Sirohi, S.C. (1994). Selection of pollinizers for mango cv. Langra. Recent Hort., 1:14-17.

Ram, S. and Sirohi, S.C., (1988). Studies on high density orcharding in mango cv. Dashehari. Acta Hort., 231: 339-44.

Ram, S. and Yadav, V.K. (1999). Mango malformation: A Review. J. Applied Hort., 1:72-78.

Ram, S. Singh, C.P. and Shukla, P. (2001). Effect of different planting densities on growth and yield of mango. India J.Hort., 58(3): 191-195.

Ram, S., Sing, C.P. and Kumar, S. (1997). A success story of high density orcharding in mango. Acta Hort., 455:375-82.

Rameswar, A. (1989). Mango flowering-stress induced. Acta Hort., 231:433-439.

Rao, V.N.M and Shanmugavelu, K.G. (1976). Studies on the effect of pruning in mango. Prog. Hort., 8: 21-25.

Reinhoudt, K (1997). Calculation on pear planting systems. Fruitteelt-Den-Haag, 87(2):14-16.

Robinson T.L. (2005). Development in high density sweet cherry pruning and training systems around the world. Acta Horticulturae 667: 269-272.

Rodriguez, W., Arya J.C. and Perez I. (2007). Effect of plant spacing and high density on canopy structure and efficiency growth and yield of banana (*Musa* AAA cv Willium). Corbana 33:1-14.

Rogers H.T. (1986). Stick with cling peaches . Western Fruit growers 106(12): 12B-12C.

Rud, G.Y.A., Tanaser, V.K. and Chimpoesh, G.P. (1976). Light relations in apple tree with palmette crown in relation to rootstocks and spacing. Trudy Kishinevskoga S. Kh. Instituta, 154:31-35.

Russell G., Jarvis P.G. and J.L. Monteith (1989). Absorption of radiation canopies and stand growth . pp21-39. , In G Russell, B. Marshell and P.G. Jarvis(eds.) Plant canopies : their growth form and function and function. Soc. Exp. Biol. Sem. Ser. 31. Cambridge University Press, Cambridge.

Sansavini, S. (1982). High density orchards of various crops. Proc. XXIst Intl. Hort. Cong., 1:182-197.

Leopold. 1974. Rep. E. Malling Res. Sta. 1946/1974: 49-54. Robitaille, H.A. 1975. J. Amer. Soc. Hort. Sci. 100: 524-27 Robitaille, H.A. and A.C.

Sansavini, S., Corelli, L. and Giunchi, L. (1984). Stress induced ethylene evolution and its possible relationship to auxin transport, cytokinin levels and flower bud induction in shoots of apple seedlings and bearing apple trees. Plant Growth Regulator, 24:127-134.

Sceicz, G. (1974). Solar radiation in crop canopies. J.Apple. Ecol., 11: 1117-1156.

Schmitz-Hubsch, H., and Furst, L. (1959). Intensiv Obstbau in Heckenform. Stuttgart.

Schaffer, B. and Gaye, G.O. (1989). Seasonal effects of pruning on light penetration, specific leaf density and chlorophyII content of mango. Scientia Hort., 41:55-61.

Shanmugavelu, K.G. and Selvarajan, M. (1985). Studies on the effect of pruning of certain varieties of mango. Second International symposium on Mango (Abst). Bangalore, pp.167.

Sharma, B.B. (1953). Studies on the disease of *Mangifera* Indica L. Proc., 4th Indian Cong., Part III. Abstracts. Sec. VI. Botany. 70-71.

Sharma, D.D. and Chauhan, J.S. (1992). Effect of different rootstocks and training systems on photosynthetic efficiency and fruit quality of Delicious apple. Indian J. Hort., 49(4): 332-37.

Sharma, J.N., Chohan, G.S., Vij, V.K. and Monga, P.K. (1992). Effect os spacing on growth , yield, and quality of kinnow mandarin. Indian J Hort., 49: 158-164.

Sharma, J.N., Josan, J.S. and Thind, S.K. (1997). Effect of pruning intensities on the productivity of kinnow mandarin. Indian J. Hort., 54(4):304-307.

Sharma, N. and Jindal, K.K. (1991). Transport and distribution of labeled auxin in geotropically stimulated fruit-bearing shoots of Royal Delicious apple (*Malus domestica* Borkh.) J. Nuclear Agric. Biol., 20:33-38.

Sharma, N. and Jindal, K.K. (1992). Effect of tree orientation on vegetative growth and spur formation in Royal Delicious apple. Indian J. Hort., 19(3):277-30.

Sharma, R.R., Singh Room and Singh Desh beer (2005). Influence of pruning intensity on light penetration and leaf physiology in high-density orchards of mango tree. Fruits 61(2): 117-123.

Sharma, Y.K., Goswami, A.M. and Sharma, R.R. (1992). Effect of dwarfing aneupliod guava rootstock in high density orcharding. Indian J. Hort., 49(1): 31-36.

Sharma. K.K; Jawanda, J.S.; Singh M.P. and Bajwa, M.S. (2004). Effect of different spacing and pruning intensities on the growth and fruit yield of Umran ber. Punjab Horticulture Journal. 20 (1/2): 47-51.

Shikhamany, S.D. (1983). Effect of the time and different doses of N and K on the growth, yield and quality of Thompson seedless grapes. Ph. D. thesis. University of Agricultural Sciences, Bangalore.

Shikhamany, S.D. (1988). Apical dominance in grapes a review. J. Maharashtra Agri. Univ., 13(1):43-48.

Shikhamany, S.D. (2001). Canopy management of the tropical and subtropical fruit crops. India J. Hort., 58(1-2): 28-32.

Shikhamany, S.d. and Chadha, K.L, (1993). Grape nutrition. In: Advances in Hort., Vot. 2. Pp. 867-878. Edited by Chadha, K.L. and Pareek, O.P., Malhotra Publishing House, New Delhi.

Shinde A.K., Waghmare G.M., Godse S.K. and Patil B.P. (2002). Pruning for rejuvenation of overcrowded, old Alphonso mango (*Mangifera* india) gardens in Konkan . Indian Journal of Agricultural Sciences 72(2): 90-92.

Bregalio S., Donatelli M., Canfalonieri R., Acutia M. (2011). Multimetric evaluation of leaf wetness models for large -area application of plant disease models. Agri. Forest Meteo 151. 1163-1172.

Sibma, L. (1977). Maximization of arable crop yields in the Nether land. Neth. J. Agr. Sci., 25:278-287.

Siham, Myriam; Bussi claude, Lescoret, Cenard Michel, Habib Robert and Gilreath James 2005. Pruning intensity and fruit load influence on vegetative and fruit growth in Alexandra peach. Pro. Fla. State Hort. Soc. 118: 266-269.

Singh S. and Singh J. 2004. Long term influence of rootstocks on growth, yield and fruit quality attributes of mango cv. Bombai. Horticultural Journal. 17(1): 15-23.

Singh G (2005). Meadow orchard system in guava production. Indian Horticulture, pp. 17-18.

Singh, Gorakh (2011). Application of canopy architecture in high density planting in guava. Progressive Horticulture 43(1): 36-43.

Singh, H.P and Chadha, K.L. (1996). Banana and Plantain in India. Infomusa, 5: 22-25.

Smart, R.E. (1973). Sunlight intercept by the Vine yards. Amer. J.E. nol. Viti, 24:141-147.

Sneath, G. (1988). Navel orange skirting trial. Farmer's News Letter Horticulture, NSW. Australia, No. 166.

Snelgar, W.P. and Thorp, T.G. (1988). Leaf area, final fruit weight and productivity in Kiwifruit. Scientia Hort., 36:241-249.

Somkuwar, R.G., Samarth Roshni, Ghule, V.S. and Sharma, A.K. (2020). Crop Load regulation to improve yield and quality of Manjari Naveen grape. Indian J. of Hort. 77(2): 381-383.

Sommer, K.J. (1995). Mechanized pruning in Australia. Proceeding of thc 4th International ATW symposium held in Stuttgart, Germany, 364:23-50.

Soudagar T.P., Haldhankar P.M., Parulekar Y.R., Dalvi V.V., and Ghule V.S.(2018). Study on the effect of tip pruning on induction of flowering and harvesting in alphonso mango. Indian J. of Hort. 75(4): 709-711.

Srivastava, A.K. and Singh, S. (2006). In Northeast India: Managing nutrient deficiencies in citrus crops. India Hort., 51(1): 6-9.

Stark, P., Jr. (1974). Amer. Fruit Grow. 94(11):12-14.

Stover, R.H.(1984). Canopy management in Valery and Grand Nain using leaf area index and photo synthetically active radiation measurements. Fruits, 329(2):89-93.

Swaroop, M., Ram, S., Singh, k C.P. and Shukla, P. (2001). Effect of pruning on growth, flowering and fruiting in mango. Indian J. Hort., 54(4):303-08.

Taylor, B.H and Ferree, D.C. (1984). The influence of summer pruning and cropping on growth and fruiting of apple. J. Amer. Soc. Hort.Sci., 109:19-24.

Teeffelen, W.V. and Teeffelen, V. (1993). Economic result determines optimal plant spacing. Fruitiest Den Haag, 83(2): 16-18.

Teeffelen, W.V. and Teeffelen, V. (1993). The influence of summer pruning and cropping on growth and fruiting of apple. J. Amer. Soc. Hort. Sci., 109:19-24.

Teskey, B.J.E. and Shoemaker, J.S. (1972). In: Tree fruit production, The AVI Publishing Co. Inc., pp.294-314.

Thippesha, D., Raju B., Jagannath, S., Naik D.J., Mahantesh B., and Poornima G. (2005). Studies on planting system and spacing on growth characters of banana cv Robusta (Musa AAA) under transitional zone of Karnataka. Karnataka Journal of Horticulture, 1:63-69.

Thomas, J.E. and Bernard, C. (1937). Fruit bud studies. III. The Sultana (Thompson Seedless). Some relation between shoot growth, chemical composition, fruit bud formation and yield. J. Can. Sci., 10: 143-57.

Tribulato, E., Continella, G., and La Rosa, G. (1992). Research on higher density planting for orange and lemon. Proc. Int. Soc. Citriculture. VII Int. Citrus Cong, Italy. 2:702-704.

Tromp, J. (1986). Growth and flower-bud formation in apple as affected by paclobutrazol, dominozide and tree orientation in combination with various gibberellins. J. Hort. Sci., 62: 433-40.

Tukey, L.D. (1982). Penn. Sta. Hort. Rev., 31:3-5.

Tumbo, S.D., Salyani, M., Witney, J.D., Wheaton, T.A. and Miller, W.M. (2002). Investigation of laser and ultrasonic ranging sensor for measurement of citrus canopy volume. Applied. Eng. Agri., 18(3):367-372.

Turner D.W. 1998. Ecophysiology of bananas: the generation of function of the leaf the leaf canopy. Acta. Hort. 490: 211-221.

Tustin S.L. Corelli-Grappadelli and G. Ravaglia. (1992). Effect of previous season and current lioght environment on early season spur development and photosynthate translocationin 'Golden Delicious' and apple. Journal of Hort.Sci. 65:351-360.

Unrath, C.R. (1999). Prohexa-dione-Calcium: A promising chemical for controlling vegetative growth of apples. Hort. Sci., 34(7):1197-1200.

Usenik V., Stempar F., Sturm K., Fajit N., (2005). Rootstocks affect leaf mineral composition and quality of Lapins sweet cherry . Acta Hort. 667: 247-252.

Usenik Valentina; Orazem Primoz and Stanpar Franci. (2010). Low leaf to fruit ratio delays fruit maturity of Lapins sweet cherry on Gisela 5. Scientia Horticulturae, 126: 33-36.

Valdivia, V.V., Garcia, S.S. and Barraza, M.H.P. (1999). 'Esmeralda' interstocks reduce 'Ataulfo' mango tree size with no reduction in yield; results of first five years. Acta Hort., 509: 291-300.

Volschenk, C.G. and Hunter, J.J. (2001). Effect of seasonal canopy management on the performance of Chenin blanc/99 Richter grapevines. South African J. Enology Viticulture, 22(1):36-40.

Volschenk, C.G. and Hunter, J.J. (2001). Effect of Trellis conversion on the performance of Chenin blanc/99 Richte grapevines. South African J. Enology Viticulture , 22(1):31-35.

Wagenmaker, P.S. (1989). High Density planting system trial with pear. Acta Hort, 243:303-307.

Wagenmakers, P.S. (1988). Planting system in relation to light. Fruitteelt .78:49.

Wang. W.C. (1987). Journal of Agricultural Research of China 36(2): 190-195.

Wareing, P.F. (1970). Growth and its coordination in trees. In: physiology Of Tree Crops. 9edited by Luckwill, L.C. and Cutting, C.V.). Academic Press, London, U.K., 1-21.

Warner, John (1991). Rootstock affects primary scaffold branch crotch angle of apple tree. Hort Science. 26(10): 1266-1267.

Wash. State Hort. Assn., 79:129-133.

Watson, D.J. (1952). The physiological basis of variation in yield. Adv. Agron., 4: 101-105.

Webster, A.D. and Shepherd, U.M. (1984). J. Hort. Sci., 59:175-182.

Wertheim S.j., P.S. WagenmakersJ and M.J. Groot 2001. Orchrdsystem and pear : conditions for success. Acta Hort. 557: 209-227.

Wertheim, S.J. (1968). The training of the slender spindle. Pub. Proesfstation Fruiteelt Wilhelmin Dadrop, Netherland No. 7.

Wertheim, S.J., de Jager, A. and Duyzens, M.J.J.P. (1986). Comparison of single row and multi-row planting system with apple, with regard to productivity, fruit size and color and light condition. Acta Hort., 160: 243-258.

Wevester, A.D. (1981). Dwarfing rootstocks for plum and cherries. Acta Hort., 114:201-207.

Whiley, A.W. and Schaffer, B. (1994). Avocado. In: Handbook of Environmental Physiology of Fruit Crops. Vol II. Subtropical and tropical Crop, pp.3-35, edited by Schaffer B. and Anderson, P.C. CRC Press, Boca Raton.

Whiting M.D. Lang G.A. (2004). 'Bing' sweet cherry on dwarfing rootstock Gisela- 5: Thinning affects fruit quality and vegetative growth but not net Co_2 exchange . J. Am. Soc. Hort. Sci. 129: 407-415.

Whiting M.D., Lang G., Ophardt D. (2005). Rootstock and training system affect sweet cherry growth , yield and fruit quality . Hort Science. 40(2): 582-586.

Wolf, T.K., Zoecklein, B.W., Cook, M.K. and Cottingham, C.K. (1990). Shoot topping and ettephon effects on White Riesling grapes and grapevines. Amer. J. Ecol. Viticulture., 41(4): 330-341.

Verner, L. (1955). Hormone relation in the growth and training of apple trees Idaho Agr. Exp. Sta. Res. Bull. 28

Wunderer, W. and Mayer, N. (1994). Comparison of some high training systems with the most important Austrain Red Wine cultivars. Mitteilungen Klosterneuburg. Rebeund-Wein Obstbau- und - Fruchteverwertung., 44(6):189-200.

Wunsche, J.N. and Lakso, A.N. (2000). Apple tree physiology-implications for orchard and tree management. Compact Fruit Tree, 33(3):82-88.

Xiong, Li., Bai Yun, Sun, Fruong, Li., Hong Ying, Miace, Li Ping, Gace, Xo, Li, By, Sun, Fr, Li., Miace, H.Y. and Gace, I.P. (1998). Effect of light distribution in the crown canopy on the yield and fruit quality of pear cultivar Pingguoli. China Fruits, 1: 23-24.

Yang, H.M., Ozaki, T., Ichii, T., Nakanishi, T. and Kawai, Y. (1992). Diffusible and extractable auxins in young Japanese pear trees. Scientia Horticulturae., 51:97-106.

Zahn F.G. (1992). Are close planting distance also possible with steinobst. Obstbau 9:430-438.

Zhakole, A.G., Negru, P.V. and Gangash, M.V.(1979). Referetivi Zhumal. 10:785.

Important Definitions

Canopy architecture-Natural expression of the genetic makeup of a tree genotypes vary in their canopy size and vigor.

Canopy management- It is a cultural practices involving skirting, hedging, pruning and re-growth management for high, quality and sustaining orcharding system.

Canopy- Refers to physical composition of tree comprising of the stem, branches shoots and leaves.

Chemical training- Decapitated shoots treated with appropriate concentration of Indole-butyric acid in lanoline resulted in wide branch angle however, prevention of down ward movement of auxin by girdling resulted in narrow-angled branch.

Cincturing- Complete removal of bark, it is usually done in late summer/early autumn. Width of cincture is 2-4 mm practiced in early maturing litchi cultivar to prevent a second late summer or autumn growth flush.

Compensation point- Minimum light intensity at which respiration rate equals photosynthetic rate, higher the temperature higher will be compensation point.

Crotch angle- Angle between the trunk and side branches. Wider the crotch angle stronger the branch, narrower the crotch angle weaker the branch.

Hedging, shearing and mowing- Refers to mechanical heading shoots at tops and sides of trees.

Hedging- Removal of the terminal portion of a shoot also known as tipping or in branch termed as stubbing.

Ideotype- Biological model of a plant frame for particular environment which may perform in a predictable manner in a specific environment.

Indiscriminate pruning- Random removal of growth by hand or mechanical hedging machine that results primarily strong re-growth.

Latitude- An angular distance measured in degrees north or south of the equator.

Leaf area index- The ratio of leaf area of a crop to the ground area on which it grows.

Meadow orchard- A type of super intensive system of orcharding where in plants are induced to form flower bud in first year and harvesting the fruit by mowing the orchard with a combine harvester hence, known meadow orchard.

Mowing- An operation consisting of trimming plant at certain height.

Multistoried Cropping (Multi-tier)- Growing of different crops having varied height, canopy orientation and rooting habits together for proper utilization of solar energy.

Propping- Timber support to hold up heavily fruiting branches to prevent fruit from touching ground.

Pruning- Judicious removal of plants parts to have balance between vegetative and reproductive growth.

Radiant flux- Amount of radiant energy per unit time emitted or received across a particular area.

Radiation- It is a process of transmission of energy by electro-magnetic waves between two bodies without media.

Saturation light intensity- Maximum light at which the rate of photosynthesis reaches a maximum value.

Skirting- Pruning of lower branches or 'skirt' of the tree to aid in weed control. Surface mulch applications, mini-sprinkler irrigation distribution and to prevent fruit of the lower branches from touching ground. A maximum height 0.5m is recommended. Repeated skirting by selective hand pruning can eventually result in strong, robust branches able to support fruit with- out touching the ground.

Solar radiation- Sun is the prime source of energy coming from the atmosphere it is source of energy for food production. Sucker-An undesirable sprout from the root or from crown.

Thinning- Removal of a shoot or branch at its point of origin.

Topping- Generally refer the non-selective mechanical hedging of tree tops

Training- Cutting away of portion of a tree to obtain the desired shape and frame-work.

Containment pruning- Pruning of the tree to contain in its allotted space

Colour Plates

Chapter 4: Canopy Development and Management of Mango

Fig. 15: Mango Variety 'Amrapali' in Full Bearing (Age 5 Years), 3x3 Distance

Fig. 16: Canopy Development in Young Orchard

Fig. 17: Amrapali form Irrgular Canopy

Fig. 18: HDP Orchard of 'Amrapali' Bearing after 4 Years

Fig. 19. Training of Young Mango Plant

Fig. 20. Twenty Five Years Mango Variety Dashehari in Bering after Rejuvenation Spacing 2.5x2.5 m.

Fig. 21.

Fig. 22: Promoting canopy development in rejuvenated HDP orchard

Fig. 23: Containing canopy spread through annual pruning

Chapter 7: Canopy Development and Management of Guava

Fig. 27. Pruning in HDP Guava

Fig. 28. Removal of 50-70% Annual Growth and Central Shoot Training

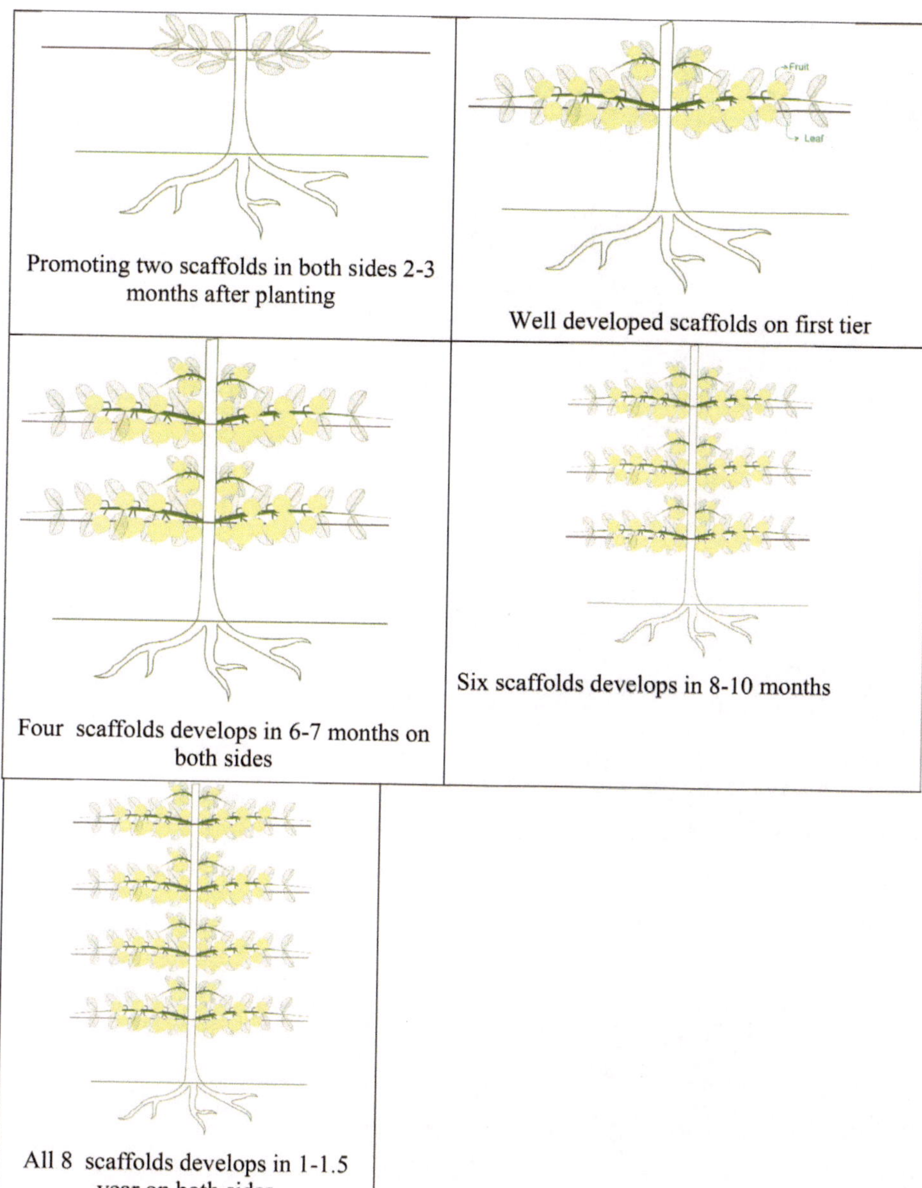

Fig. 29. Development of espalier architecture

Fig. 30. Tree laden with fruits during first year

Fig. 31. Tree laden with winter crop

All 8 scaffolds develops in 1-1.5 m on both sides

Fig. 32. Espalier architecture system

Chapter 9: Canopy Development and Management of Apple

Fig. 33. Apple on espalier architecture

Fig. 34. Spur type

Fig. 35. Apple variety Red Spur

Fig. 36. Apple on single axis training system

Fig. 37. Apple var Starking Delicious

Fig. 38. Apple var Red Spur in Y shape architecture